Fundamentals of Quantum Chemistry

Fundamentals of Quantum Chemistry

Molecular Spectroscopy and Modern Electronic Structure Computations

Michael Mueller

Rose-Hullman Institute of Technology
Terre Haute, Indiana

Kluwer Academic / Plenum Publishers
New York Boston Dordrecht London Moscow

ISBN: 0-306-46596-5

233 Spring Street, New York, New York 10013

http://www.wkap.nl

10 9 8 7 6 5 4 3 2 1

A C.I.P. record for this book is available from the Library of Congress

Printed in the United States of America

Foreword

As quantum theory enters its second century, it is fitting to examine just how far it has come as a tool for the chemist. Beginning with Max Planck's agonizing conclusion in 1900 that linked energy emission in discreet bundles to the resultant black-body radiation curve, a body of knowledge has developed with profound consequences in our ability to understand nature. In the early years, quantum theory was the providence of physicists and certain breeds of physical chemists. While physicists honed and refined the theory and studied atoms and their component systems, physical chemists began the foray into the study of larger, molecular systems. Quantum theory predictions of these systems were first verified through experimental spectroscopic studies in the electromagnetic spectrum (microwave, infrared and ultraviolet/visible), and, later, by nuclear magnetic resonance (NMR) spectroscopy.

Over two generations these studies were hampered by two major drawbacks: lack of resolution of spectroscopic data, and the complexity of calculations. *This powerful theory that promised understanding of the fundamental nature of molecules faced formidable challenges.* The following example may put things in perspective for today's chemistry faculty, college seniors or graduate students: As little as 40 years ago, force field calculations on a molecule as simple as ketene was a four to five year dissertation project. The calculations were carried out utilizing the best mainframe computers in attempts to match fundamental frequencies to experimental values measured with a resolution of five to ten wavenumbers in the low infrared region! Post World War II advances in instrumentation, particularly the spin-offs of the National Aeronautics and Space Administration (NASA) efforts, quickly changed the landscape of high-resolution spectroscopic data. Laser sources and Fourier transform spectroscopy are two notable advances, and these began to appear in undergraduate laboratories in the mid-1980s. At that time, only chemists with access to supercomputers were to realize the full fruits of quantum theory. This past decade's advent of commercially available quantum mechanical calculation packages, which run on surprisingly sophisticated laptop computers, provide approximation technology for all chemists. Approximation techniques developed by the early pioneers can now be carried out to as many iterations as necessary to produce meaningful results for sophomore organic chemistry students, graduate students, endowed chair professors, and pharmaceutical researchers. The impact of quantum mechanical calculations is also being felt in certain areas of the biological sciences, as illustrated in the results of conformational studies of biologically active molecules. Today's growth of quantum chemistry literature is as fast as that of NMR studies in the 1960s.

An excellent example of the introduction of quantum chemistry calculations in the undergraduate curriculum is found at the author's institution. Sophomore organic chemistry students are introduced to the PC-Spartan+® program to calculate the lowest energy of possible structures. The same program is utilized in physical chemistry to compute the potential energy surface of the reaction coordinate in simple reactions. Biochemistry students take advantage of calculations to elucidate the pathways to creation of designer drugs. This hands-on approach to quantum chemistry calculations is not unique to that institution. However, the flavor of the department's philosophy ties in quite nicely with the tone of this textbook that is pitched at just the proper level, advanced undergraduates and first year graduate students.

Farrell Brown
Professor Emeritus of Chemistry
Clemson University

Preface

This text is designed as a practical introduction to quantum chemistry for undergraduate and graduate students. The text requires a student to have completed a year of calculus, a physics course in mechanics, and a minimum of a year of chemistry. Since the text does not require an extensive background in chemistry, it is applicable to a wide variety of students with the aforementioned background; however, the primary target of this text is for undergraduate chemistry majors.

The text provides students with a strong foundation in the principles, formulations, and applications of quantum mechanics in chemistry. For some students, this is a terminal course in quantum chemistry providing them with a basic introduction to quantum theory and problem solving techniques along with the skills to do electronic structure calculations - an application that is becoming increasingly more prevalent in all disciplines of chemistry. For students who will take more advanced courses in quantum chemistry in either their undergraduate or graduate program, this text will provide a solid foundation that they can build further knowledge from.

Early in the text, the fundamentals of quantum mechanics are established. This is done in a way so that students see the relevance of quantum mechanics to chemistry throughout the development of quantum theory through special boxes entitled *Chemical Connection*. The questions in these boxes provide an excellent basis for discussion in or out of the classroom while providing the student with insight as to how these concepts will be used later in the text when chemical models are actually developed.

Students are also guided into thinking "quantum mechanically" early in the text through conceptual questions in boxes entitled *Points of Further Understanding*. Like the questions in the *Chemical Connection* boxes, these questions provide an excellent basis for discussion in or out of the classroom. These questions move students from just focusing on the rigorous mathematical derivations and help them begin to visualize the implications of quantum mechanics.

Rotational and vibrational spectroscopy of molecules is discussed in the text as early as possible to provide an application of quantum mechanics to chemistry using model problems developed previously. Spectroscopy provides for a means of demonstrating how quantum mechanics can be used to explain and predict experimental observation.

The last chapter of the text focuses on the understanding and the approach to doing modern day electronic structure computations of molecules. These types of computations have become invaluable tools in modern theoretical and experimental chemical research. The computational methods are discussed along with the results compared to experiment when possible to aide in making sound decisions as to what type of Hamiltonian and basis set that should be used, and it provides a basis for using computational strategies based on desired reliability to make computations as efficient as possible.

There are many people to thank in the development of this text, far too many to list individually here. A special thanks goes out to the students over the years who have helped shape the approach used in this text based on what has helped them learn and develop interest in the subject.

Terre Haute, IN Michael R. Mueller

Acknowledgments

Farrell B. Brown Clemson University

Rita K. Hessley University of Cleveland College of Applied Science

Daniel L. Morris, Jr. Rose-Hulman Institute of Technology

Gerome F. Wagner Rose-Hulman Institute of Technology

The permission of the copyright holder, Prentice-Hall, to reproduce Figure 7-1 is gratefully acknowledged.

The permission of the copyright holder, Wavefunction, Inc., to reproduce the data on molecular electronic structure computations in Chapter 9 is gratefully acknowledged.

Contents

Chapter 1

Classical Mechanics

Classical mechanics arises from our observation of matter in the macroscopic world. From these everyday observations, the definition of particles is formulated. In classical mechanics, a particle has a specific location in space that can be defined precisely limited only by the uncertainty of the measurement instruments used. If all of the forces acting on the particle are accounted for, an exact energy and trajectory for the particle can be determined. Classical mechanics yields results consistent with experiment on macroscopic particles; hence, any theory such as quantum mechanics must yield classical results at these limits.

There are a number of different techniques used to solve classical mechanical systems that include Newtonian and Hamiltonian mechanics. Hamiltonian mechanics, though originally developed for classical systems, has a framework that is particularly useful in quantum mechanics.

1.1 NEWTONIAN MECHANICS

In the mechanics of Sir Isaac Newton, the equations of motion are obtained from one of Newton's Laws of Motion: Change of motion is proportional to the applied force and takes place in the direction of the force. Force, $\vec{F}$, is a vector that is equal to the mass of the particle, m, multiplied by the acceleration vector $\vec{a}$.

$$\vec{F} = m\vec{a} \tag{1-1}$$

If the resultant force acting on the particle is known, then the equation of motion (i.e. trajectory) for the particle can be obtained. The acceleration is the second time derivative of position, q, which is represented as

$$\bar{a} = \frac{d^2 q}{dt^2} = \ddot{q} . \tag{1-2}$$

The symbol q is used as a general symbol for position expressed in any inertial coordinate system such as Cartesian, polar, or spherical. A double dot on top of a symbol, such as $\ddot{q}$, represents the second derivative with respect to time, and a single dot over a symbol represents the first derivative with respect to time.

$$\dot{q} \equiv \frac{dq}{dt}$$

The systems considered, until later in the text, will be ***conservative*** systems, and masses will be considered to be point masses. If a force is a function of position only (i.e. no time dependence), then the force is said to be conservative. In conservative systems, the sum of the kinetic and potential energy remains constant throughout the motion. ***Non-conservative*** systems, that is, those for which the force has time dependence, are usually of a dissipation type, such as friction or air resistance. Masses will be assumed to have no volume but exist at a given point in space.

Example 1-1

Problem: Determine the trajectory of a projectile fired from a cannon whereby the muzzle is at an angle α from the horizontal x-axis and leaves the muzzle with a velocity of v_m. Assume that there is no air resistance.

Solution: This problem is an example of a ***separable problem***: the equations of motion can be solved independently in the horizontal and vertical coordinates. First the forces acting on the particle must be obtained in the two independent coordinates.

Horizontal axis (x-axis): $F_x = m\ddot{x} = 0$

Vertical axis (y-axis): $F_y = m\ddot{y} = -mg$

The forces generate two differential equations to be solved. Upon integration, this results in the following trajectories for the particle along the x and y-axes:

$$\ddot{x} = 0; \qquad x(t) = x_0 + (v_m \cos\alpha)t$$

$$\ddot{y} = -g; \qquad y(t) = y_0 + (v_m \sin\alpha)t - \tfrac{1}{2}gt^2$$

The constant x_0 and y_0 represent the projectile at the origin (i.e. initial time).

1.2 HAMILTONIAN MECHANICS

An alternative approach to solving mechanical problems that makes some problems more tractable was first introduced in 1834 by the Scottish mathematician Sir William R. Hamilton. In this approach, the Hamiltonian, H, is obtained from the kinetic energy, T, and the potential energy, V, of the particles in a conservative system.

$$H = T + V \tag{1-3}$$

The kinetic energy is expressed as the dot product of the momentum vector, $\vec{p}$, divided by two times the mass of each particle in the system.

$$T = \sum \frac{\vec{p} \cdot \vec{p}}{2m} = \sum \frac{p^2}{2m} \tag{1-4}$$

The potential energy of the particles will depend on the positions of the particles. Hamilton determined that for a generalized coordinate system, the equations of motion could be obtained from the Hamiltonian and from the following identities:

$$\left(\frac{\partial H}{\partial q}\right)_p = -\frac{dp}{dt} = -\dot{p}\,, \tag{1-5}$$

and

$$\left(\frac{\partial H}{\partial p}\right)_q = \frac{dq}{dt} = \dot{q}\,. \tag{1-6}$$

Simultaneous solution of these differential equations through all of the coordinates in the system will result in the trajectories for the particles.

Example 1-2

Problem: Solve the same problem as shown in Example 1-1 using Hamiltonian mechanics.

Solution: The first step is to determine the Hamiltonian for the problem. The problem is still separable and the projectile will have kinetic energy in both the x and y-axes. The potential energy of the particle is due to gravitational potential energy given as V = mgy.

$$H(x, y, p_x, p_y) = T + V = \left(\frac{{p_x}^2}{2m} + \frac{{p_y}^2}{2m}\right) + mgy$$

Now the Hamilton identities in Equations 1-5 and 1-6 must be determined for this system.

$$\left(\frac{\partial H}{\partial x}\right)_{p_x, p_y, y} = 0 = -\dot{p}_x \qquad \left(\frac{\partial H}{\partial y}\right)_{p_x, p_y, x} = mg = -\dot{p}_y$$

$$\left(\frac{\partial H}{\partial p_x}\right)_{p_y, x, y} = \frac{p_x}{m} = \dot{x} \qquad \left(\frac{\partial H}{\partial p_y}\right)_{p_x, x, y} = \frac{p_y}{m} = \dot{y}$$

The above formulations result in two non-trivial differential equations that are the same as obtained in Example 1-1 using Newtonian mechanics.

$\ddot{x} = 0$	$\ddot{y} = -g$

This will result in the same trajectory as obtained in Example 1-1.

$x(t) = x_0 + (v_m \cos\alpha)t$	$y(t) = y_0 + (v_m \sin\alpha)t - \frac{1}{2}gt^2$

Notice that in Hamiltonian mechanics, initially the momentum of the particles is treated separately from the position of the particles. This method of treating the momentum separately from position will prove useful in quantum mechanics.

1.3 THE HARMONIC OSCILLATOR

The harmonic oscillator is an important model problem in chemical systems to describe the oscillatory (vibrational) motion along the bonds between the atoms in a molecule. In this model, the bond is viewed as a spring with a force constant of k.

Consider a spring with a force constant k such that one end of the spring is attached to an immovable object such as a wall and the other is attached to a mass, m (see Figure 1-1). Hamiltonian mechanics will be used; hence, the first step is to determine the Hamiltonian for the problem. The mass is confined to the x-axis and will have both kinetic and potential energy. The potential energy is the square of the distance the spring is displaced from its equilibrium position, x_0, times one-half of the spring force constant, k (Hooke's Law).

$$H = \frac{p_x^2}{2m} + \frac{1}{2}k(x - x_0)^2 \qquad (1\text{-}7)$$

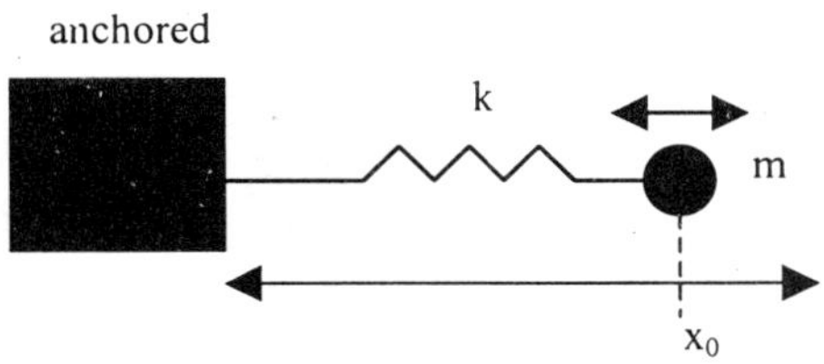

Figure 1-1. The harmonic spring is attached to an immovable object at one end and an oscillating mass m at the other end. The constant x_0 corresponds to the point of zero potential energy. Assume the only force acting on the mass m is along the horizontal axis.

Taking the derivative of the Hamiltonian (Equation 1-7) with respect to position and applying Equation 1-5 yields:

$$\left(\frac{\partial H}{\partial x}\right)_p = k(x - x_0) = -\dot{p} .$$

Taking the derivative of the Hamiltonian (Equation 1-7) with respect to momentum and applying Equation 1-6 yields:

$$\left(\frac{\partial H}{\partial p}\right)_x = \frac{p}{m} = \dot{x} .$$

The second differential equation yields a trivial result:

$$\frac{p}{m} = \dot{x} = \dot{x} ;$$

however, the first differential equation can be used to determine the trajectory of the mass m. The time derivative of momentum is equivalent to the force, or mass times acceleration.

$$\frac{dp}{dt} = \frac{d}{dt}\left(m\frac{dx}{dt}\right) = m\frac{d^2x}{dt^2} = -k(x - x_0) \qquad \text{(1-8a)}$$

or

$$\ddot{x} = -\left(\frac{k}{m}\right)(x - x_0) \qquad \text{(1-8b)}$$

The solution to this differential equation is well known. One solution is given below.

$$x(t) = x_0 + a\sin(\omega t) + b\cos(\omega t) \qquad \text{(1-9)}$$

Another mathematically equivalent solution can be found by utilizing the following Euler identities ($i = \sqrt{-1}$):

$$e^{ix} = \cos x + i\sin x \qquad \text{(1-10a)}$$

and

$$e^{-ix} = \cos x - i\sin x\,. \qquad \text{(1-10b)}$$

This results in the following mathematically equivalent trajectory as in Equation 1-9:

$$x(t) = x_0 + Ae^{i\omega t} + Be^{-i\omega t}\,. \qquad \text{(1-11)}$$

The value of x_0 is the equilibrium length of the spring. Since the product of ωt must be dimensionless, the constant ω must have units of inverse time and must be the frequency of oscillation. By taking the second time derivative of either Equation 1-9 or 1-11 results in the following expression:

$$\ddot{x} = -\omega^2(x(t) - x_0)\,. \qquad \text{(1-12)}$$

By comparing Equation 1-12 with Equation 1-8b, an expression for ω is readily obtained.

$$\omega = \sqrt{\frac{k}{m}} \tag{1-13}$$

Since the sine and cosine functions will oscillate from +1 to –1, the constants a and b in Equation 1-9 and likewise the constants A and B in Equation 1-11 are related to the amplitude and phase of motion of the mass. There are no constraints on the values of these constants, and the system is not quantized.

Chemical Connection

A diatomic molecule approximates the model just discussed such that the mass of one atom is much larger than the other atom: such as hydrogen bromide. In infrared spectroscopy, the absorbed infrared radiation results in transitions in both the vibrational and rotational states of a molecule. Considering only the vibrational transitions, what would an infrared spectrum of hydrogen bromide look like based on the classical result? According to classical mechanics, would infrared spectroscopy be a useful tool in chemistry?

A model can now be developed that more accurately describes a diatomic molecule. Consider two masses, m_1 and m_2, separated by a spring with a force constant k and an equilibrium length of x_0 as shown in Figure 1-2. The Hamiltonian is shown below.

$$H(x_1, x_2, p_1, p_2) = \frac{p_1^{\ 2}}{2m_1} + \frac{p_2^{\ 2}}{2m_2} + \frac{1}{2}k(x_2 - x_1 - x_0)^2$$

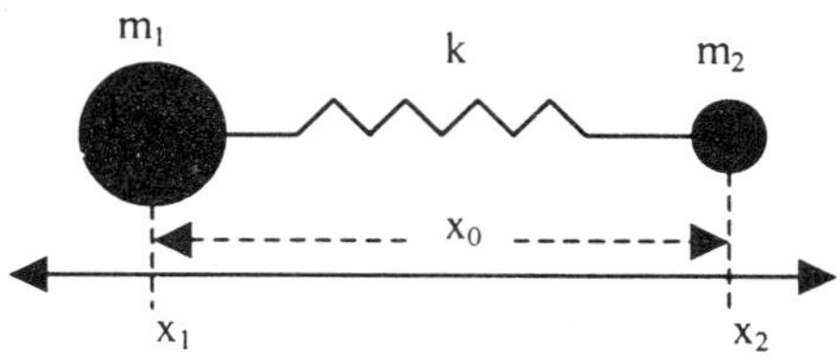

Figure 1-2. Two masses are separated by a harmonic spring with a force constant k. The particles are confined to the x-axis. The positions of the particles are designated as x_1 and x_2 with x_0 corresponding to the equilibrium spring length.

Note that the Hamiltonian appears to be inseparable. Making a coordinate transformation to a center-of-mass coordinate system can make this problem separable. Define r as the displacement of the spring from its equilibrium position and s as the position of the center of mass.

$$r \equiv x_2 - x_1 - x_0$$

$$s \equiv \frac{(m_1 x_1 + m_2 x_2)}{(m_1 + m_2)}$$

As a result of the coordinate transformation, the potential energy for the system becomes:

$$V = \tfrac{1}{2} k r^2 .$$

Now the momentum p_1 and p_2 must be transformed to the momentum in the s and r coordinates. The time derivatives of r and s must be taken and related to the time derivatives of x_1 and x_2.

$$\dot{r} = \dot{x}_2 - \dot{x}_1 \qquad (1\text{-}14)$$

$$\dot{s} = \frac{(m_1\dot{x}_1 + m_2\dot{x}_2)}{(m_1 + m_2)} \tag{1-15}$$

From Equations 1-14 and 1-15, expressions for $\dot{x}_1$ and $\dot{x}_2$ in terms of $\dot{s}$ and $\dot{r}$ can be obtained.

$$\dot{x}_1 = \dot{s} - \left[\frac{m_2}{m_1 + m_2}\right]\dot{r} \tag{1-16}$$

$$\dot{x}_2 = \dot{s} + \left[\frac{m_1}{m_1 + m_2}\right]\dot{r} \tag{1-17}$$

The momentum terms p_1 and p_2 are now expressed in terms of the center of mass coordinates s and r.

$$p_1 = m_1\dot{x}_1 = m_1\dot{s} - \left[\frac{m_1 m_2}{m_1 + m_2}\right]\dot{r}$$

$$p_2 = m_2\dot{x}_2 = m_2\dot{s} + \left[\frac{m_1 m_2}{m_1 + m_2}\right]\dot{r}$$

The reduced mass of the system, μ, is defined as

$$\mu = \frac{m_1 m_2}{m_1 + m_2}.$$

This reduces the expressions for p_1 and p_2 to the following:

$$p_1 = m_1\dot{s} - \mu\dot{r}$$

and

$$p_2 = m_2\dot{s} + \mu\dot{r}.$$

The Hamiltonian can now be written in terms of the center-of-mass coordinate system.

$$H(r,s,p_r,p_s)=\frac{p_1^{\,2}}{2m_1}+\frac{p_2^{\,2}}{2m_2}+\frac{1}{2}kr^2=\frac{1}{2}\left[(m_1+m_2)\dot{s}^2+\mu\dot{r}^2\right]+\frac{1}{2}kr^2$$

A further simplification can be made to the Hamiltonian by recognizing that the total mass of the system, M, is the sum of m_1 and m_2 (i.e. $M = m_1 + m_2$).

$$H(r,s,p_r,p_s)=\frac{1}{2}\left[M\dot{s}^2+\mu\dot{r}^2\right]+\frac{1}{2}kr^2=\frac{p_r^{\,2}}{2\mu}+\frac{p_s^{\,2}}{2M}+\frac{1}{2}kr^2 \quad (1\text{-}18)$$

Recall that the coordinate s corresponds to the center of mass of the system whereas the coordinate r corresponds to the displacement of the spring. This ensures that r and s are separable. It can be concluded that the kinetic energy term

$$\frac{p_s^2}{2m}$$

must correspond to the translation of the entire system in space. Since it is the vibrational motion that is of interest, the kinetic term for the translation of the system can be neglected in the Hamiltonian. The resulting Hamiltonian that corresponds to the vibrational motion is as follows:

$$H(r,p_r)=\frac{p_r^{\,2}}{2\mu}+\frac{1}{2}kr^2 \quad (1\text{-}19)$$

Notice that the Hamiltonian in Equation 1-19 is identical in form to the Hamiltonian in Equation 1-7 solved previously. The solution can be inferred from the previous result recognizing that when the spring is in its equilibrium position x_0, then $r = 0$ (refer to Equation 1-14).

$$r(t)=a\sin(\omega t)+b\cos(\omega t)=Ae^{i\omega t}+Be^{-i\omega t} \quad (1\text{-}20)$$

$$\omega = \sqrt{\frac{k}{\mu}} \tag{1-21}$$

This example demonstrates a number of important techniques in solving mechanical problems. A mechanical problem can at times be made separable by an appropriate coordinate transformation. This will prove especially useful in solving problems that involve circular motion where coordinates can be made separable by transforming Cartesian coordinates to polar or spherical coordinates. Another more subtle point is to learn to recognize a Hamiltonian to which you know the solution. In chemical systems, the Hamiltonian of a molecule will often have components similar to other molecules or model problems for which the solution is known. The ability to recognize these components will prove important to solving many of these systems.

PROBLEMS AND EXERCISES

1.1) Calculate the range of a projectile with a mass of 10.0 kg fired from a cannon at an angle of 30.0° from the horizontal axis with a muzzle velocity of 10.0 m/s.

1.2) Using Hamiltonian mechanics, determine the time it will take a 1.00 kg block initially at rest to slide down a 1.00 m long frictionless ramp that has an angle of 45.0° from the horizontal axis.

1.3) Set up the Hamiltonian for a particle with a mass m that is free to move in the x, y, and z-coordinates that experiences the gravitational potential $V(x,y,z) = mgy$. Using Equations 1-5 and 1-6, obtain the equations of motion in each dimension.

1.4) Determine the force constant of a harmonic spring oscillating at 50 sec^{-1} that is attached to an immovable object at one end the following masses at the other end: a) 0.100 kg; b) 1.00 kg; c) 10.0 kg; and d) 100. kg.

1.5) Determine the oscillation frequency of a $^{14}N^{16}O$ bond that has a force constant of 1607 kg sec^{-2}.

1.6) Show that a potential of the general form $V(x) = a + bx + cx^2$ is the same as that for a harmonic oscillator because it can be written as $V(x) = V_0 + \frac{1}{2}k(x - x_0)^2$. Find k, V_0, and x_0 in terms of a, b, and c.

Chapter 2

Fundamentals of Quantum Mechanics

Classical mechanics, introduced in the last chapter, is inadequate for describing systems composed of small particles such as electrons, atoms, and molecules. What is missing from classical mechanics is the description of wavelike properties of matter that predominates with small particles. Quantum mechanics takes into account the wavelike properties of matter when solving mechanical problems. The mathematics and laws of quantum mechanics that must be used to explain wavelike properties cause a dramatic change in the way mechanical problems must be solved. In classical mechanics, the mathematics can be directly correlated to physically measurable properties such as force, momentum, and position. In quantum mechanics, the mathematics that yields physically measurable properties is obtained from mathematical operations with an indirect physical correlation.

2.1 THE DE BROGLIE RELATION

At the beginning of the 20th century, experimentation revealed that electromagnetic radiation has particle-like properties (as an example, photons were shown to be deflected by gravitational fields), and as a result, it was theorized that all particles must also have wavelike properties. The idea that particles have wavelike properties resulted from the observation that a monoenergetic beam of electrons could be diffracted in the same way a monochromatic beam of light can be diffracted. The diffraction of light is a result of its wave character; hence, there must be an abstract type of wave character associated with small particles. De Broglie summarized the universal *duality* of particles and waves in 1924 and proposed that all matter has an associated wave with a wavelength, λ, that is inversely proportional to the momentum, p, of the particle (verified experimentally in 1927 by Davison and Germer).

$$p = \frac{h}{\lambda} \tag{2-1}$$

The constant of proportionality, h, is Planck's constant. The de Broglie relation fuses the ideas of particle-like properties (i.e. momentum) with wave-like properties (i.e. wavelength). This *duality of particle and wave properties* will be the theme throughout the rest of the text.

The de Broglie relationship not only provides for a mathematical relationship for the duality of particles and waves, but it also begins to hint at the idea of quantization in mechanics. If a particle is in an orbit, the only allowed radii and momenta are those where the waves associated with the particle will interfere non-destructively as they wrap around each orbit. Momenta and radii where the waves destructively interfere with one another are not allowed, as this would suggest an "annihilation" of the particle as it orbits through successive revolutions.

As mentioned in the introduction to Chapter 1, for any theory to be valid it must predict classical mechanics at the limit of macroscopic particles (called the ***Correspondence Principle***). In the de Broglie relationship, the wavelength is an indication of the degree of wave-like properties. Consider an automobile that has a mass of 1000. kg travelling at a speed of 50.0 km hr^{-1}. The momentum of the automobile is

$$p = (1000.kg)(50.0km/hr)(10^3 m/km)(1hr/3600s) = 1.39x10^4 kg \cdot m \cdot s^{-1}.$$

Dividing this result into Planck's constant yields the de Broglie wavelength.

$$\lambda = \frac{6.63x10^{-34} kg \cdot m^2 \cdot s^{-1}}{1.39x10^4 kg \cdot m \cdot s^{-1}} = 4.77x10^{-38} m$$

Considering the dimensions of an automobile, this wavelength would be beyond the accuracy of the best measuring instruments. If an electron were travelling at a speed of 50.0 km/hr, the corresponding de Broglie wavelength would be

$$\lambda = \frac{6.63x10^{-34}\, kg \cdot m^2 \cdot s^{-1}}{1.27x10^{-29}\, kg \cdot m \cdot s^{-1}} = 5.22x10^{-5}\, m\,.$$

This wavelength is quite significant compared to the average radius of a hydrogen ground-state orbital (1s) of approximately 10^{-11}m. The wave-like properties in our macroscopic world do not disappear, but rather they become insignificant. The wave-like properties of particles at the atomic scale (i.e. small mass) become quite significant and cannot be neglected. The magnitude of Plank's constant (6.63 x 10^{-34} J·s) is so small that only for very small masses is the de Broglie wavelength significant.

2.2 ACCOUNTING FOR WAVE CHARACTER IN MECHANICAL SYSTEMS

The de Broglie relationship suggests that in order to obtain a full mechanical description of a free particle (a free particle has no forces acting on it), there must be a wavelength and hence some simple *oscillating function* associated with the particle's description. This function can be a sine, cosine, or, equivalently, a complex exponential function‡.

$$A(x) = A_0 \sin\left(\frac{2\pi x}{\lambda}\right) \tag{2-2}$$

In the wave equation above, A_0 represents the amplitude of the wave and λ represents the de Broglie wavelength. Note that when the second derivative

‡ The complex exponential function e^{ikx} and e^{-ikx} (where k = 2π/λ in this case) are related to sine and cosine functions as shown in the following mathematical identities (see Equations 1-10a and 1-10b):

$$e^{ikx} = \cos(kx) + i\sin(kx) \qquad e^{-ikx} = \cos(kx) - i\sin(kx).$$

Expressing a wavefunction in terms of a complex exponential can be useful in some cases as will be shown later in the text.

of the equation is taken, the same function along with a constant, C, results.

$$\frac{d^2}{dx^2}[A(x)] = -\left(\frac{2\pi}{\lambda}\right)^2 A_0 \sin\left(\frac{2\pi x}{\lambda}\right) = CA(x) \qquad (2\text{-}3)$$

In such a situation, the function is called an ***eigenfunction***, and the constant is called an ***eigenvalue***. The eigenfunction is a wavefunction and is generally given the symbol, ψ.

What is needed now is a physical connection to the mathematics described so far. If the negative of the square of $\hbar$ ($\hbar = h/2\pi$, where h is Planck's constant) is multiplied through Equation 2-3, the square of the momentum of the particle is obtained as described in the de Broglie relation given in Equation 2-1.

$$-\hbar^2\left(\frac{d^2}{dx^2}\right)A(x) = \hbar^2\left(\frac{2\pi}{\lambda}\right)^2 A_0 \sin\left(\frac{2\pi x}{\lambda}\right) = \left(\frac{h}{\lambda}\right)^2 A(x) = p^2 A(x) \quad (2\text{-}4)$$

Equation 2-4 demonstrates a very important result that lies at the heart of quantum mechanics. When certain ***operators*** (in this case taking the second derivative with respect to position multiplied by $-\hbar^2$) are applied to the wavefunction that describes the system, an ***observable*** (in this case the square of the momentum) is obtained.

This leads to the following postulates of quantum mechanics.

Postulate 1: *For every quantum mechanical system, there exists a wavefunction that contains a full mechanical description of the system.*

Postulate 2: *For every experimentally observable variable such as momentum, energy, or, position there is an associated mathematical operator.*

Postulate 2 requires that every experimentally observable quantity have a mathematical operation associated with it that is applied to the eigenfunction of the system. Operators are signified with a "^" over the quantity. Some of the most common operators that result in observables for a system are given in the following list.

Position: $$\hat{q} = q \tag{2-5}$$

Momentum: $$\hat{p} = \left(\frac{\hbar}{i}\right)\frac{\partial}{\partial q} \tag{2-6}$$

Kinetic Energy: $$\hat{T} = -\left(\frac{\hbar^2}{2m}\right)\left(\frac{\partial^2}{\partial q^2}\right) \tag{2-7}$$

Postulates 1 and 2 lead to Postulate 3 (the Schroedinger equation) in which the Hamiltonian operator ($\hat{H} = \hat{T} + \hat{V}$) applied to the wavefunction of the system yields the energy, E, of the system and the wavefunction.

$$\hat{H}\psi = (\hat{T} + \hat{V})\psi = T\psi + \hat{V}\psi = E\psi \tag{2-8}$$

Postulate 3: *The wavefunction of the system must be an eigenfunction of the Hamiltonian operator.*

Postulate 3 requires that the wavefunction for the system to be an eigenfunction of one specific operator, the Hamiltonian. Solving the Schroedinger equation is central to solving all quantum mechanical problems.

2.3 THE BORN INTERPRETATION

So far a model has been developed to obtain the energy of the system (an experimentally determinable property – i.e. an observable) by applying an operator, the Hamiltonian, to the wavefunction for the system. This approach is analogous to how the energy of a classical standing wave is obtained. The second derivative with respect to position is taken of the function describing the classical standing wave.

$$\frac{d^2\psi(x)}{dx^2} = -\left(\frac{2\pi}{\lambda}\right)^2 \psi(x) \qquad \textit{(classical standing wave)}$$

The major difference between the quantum mechanical approach for describing particles and that of classical mechanics describing standing waves is that in classical mechanics the operator (taking the second derivative with respect to position) is applied to a function that is physically observable. At this point, the wavefunction describing the particle has no observable property beyond the de Broglie wavelength.

The physical connection of the wavefunction, ψ, must still be determined. The basis for the interpretation of ψ comes from a suggestion made by Max Born in 1926 that ψ corresponds to the *square root of the probability density*: the square root of the probability of finding a particle per unit volume. The wavefunction, however, may be a complex function. As an example for a given state n,

$$\psi_n = Ae^{ikx}.$$

The square of this function will result in a complex value. To ensure that the probability density has a real value, the probability density is obtained by multiplying the wavefunction by the complex conjugate of the wavefunction, ψ_n^*. The complex conjugate is obtained by replacing any "i" in the function with a "-i". The complex conjugate of the function above is

$$\psi_n^* = Ae^{-ikx}.$$

Consider a 1-dimensional system where a particle is free to be found anywhere on a line in the x-axis. Divide the line into infinitesimal segments of length dx. The probability that the particle is between x and x + dx is $\psi_n^*\psi_n dx$. It is important to note that $\psi_n^*\psi_n$ is not a probability but rather it is a *probability density* (i.e. probability per unit volume). To find the probability, the product $\psi_n^*\psi_n$ must be multiplied by a volume element (in the case of a 1-dimensional system, the volume element is just dx).

Born's interpretation of ψ was made from an analogy of Einstein's correlation of the number of photons in a light beam relative to its intensity. The intensity of a light beam is the sum of the square of the amplitudes of the magnetic and electric fields. Born made an analogy that the square of the wavefunction relates to the "intensity" of finding a particle in a unit volume. This analogy is accepted because it agrees well with experimental results.

The Born interpretation leads to a number of important implications on the wavefunction. The function must be single-valued: it would not make physical sense that the particle had two different probabilities in the same region of space. The sum of the probabilities of finding a particle within each segment of space in the universe ($\psi_n{}^*\psi_n$ times a volume element, $d\tau$) must be equal to unity. The mathematical operation of ensuring that the sum overall space results in unity is referred to as *normalizing the wavefunction.*

Normalization: $$\int \psi_n^* \psi d\tau = 1 \qquad (2\text{-}9)$$

The normalization condition of the wavefunction further implies that the wavefunction cannot become infinite over a finite region of space.

2.4 PARTICLE-IN-A-BOX

An instructive model problem in quantum mechanics is one in which a particle of mass m is confined to a one-dimensional box as shown in Figure 2-1. The particle is confined to the box because at the walls the potential is infinite. The potential energy inside the box is zero.

$$V(x) = 0: \qquad 0 < x < L \qquad (2\text{-}10a)$$

$$V(x) = \infty: \qquad x \leq 0 \text{ and } x \geq L \qquad (2\text{-}10b)$$

This means that the particle will have only a kinetic energy term in the Hamiltonian operator.

$$\hat{H} = \hat{T} = -\frac{\hbar^2}{2m}\left(\frac{d^2}{dx^2}\right)$$

The Schroedinger equation can now be written for the problem.

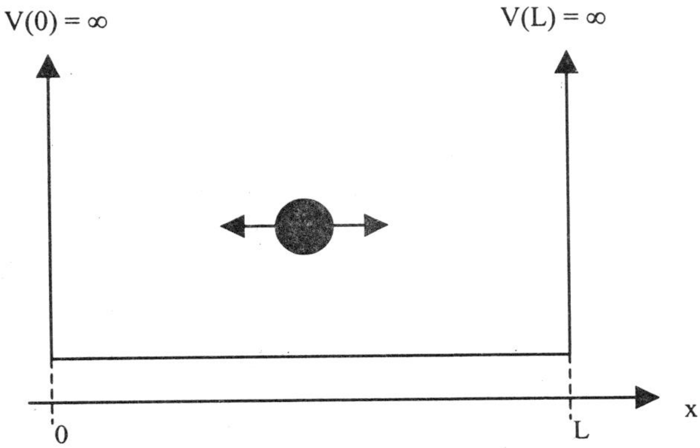

Figure 2-1. A particle confined to a one-dimensional box. The potential energy is zero within the box, and it is infinite at the walls. As a result, the particle must have zero probability at $x \leq 0$ and at $x \geq L$

$$\hat{H}\psi = -\frac{\hbar^2}{2m}\left(\frac{d^2\psi}{dx^2}\right) = E\psi \tag{2-11}$$

In order for the wavefunction, ψ, for this system to be an eigenfunction of the Hamiltonian, ψ must be a function such that taking its second derivative yields the same function. Possible functions include sine, cosine, or the mathematically equivalent complex exponential (see the footnote on page 16).

$$\psi(x) = A\sin(kx) + B\cos(kx) \tag{2-12a}$$

$$\psi(x) = Ce^{ikx} + De^{-ikx} \tag{2-12b}$$

The constants A, B, C, and D are evaluated using the boundary conditions and the normalization condition. The constant k is the frequency of the wavefunctions (frequency in the sense of inverse distance) and is also determined by the boundary conditions. If Equation 2-12a is used in the Schroedinger equation, the energy of the system is obtained in terms of k.

$$\hat{H}(x)\psi(x) = -\left(\frac{\hbar^2}{2m}\right)\left(\frac{d^2}{dx^2}\right)\left(A\sin(kx) + B\cos(kx)\right) = \left(\frac{\hbar^2 k^2}{2m}\right)\psi(x) \quad (2\text{-}13)$$

$$E = \frac{\hbar^2 k^2}{2m} \quad (2\text{-}14)$$

To determine the constant k, the boundary conditions to the problem must be applied. Recall that $\psi^*\psi$ is the probability density of the particle. The particle cannot exist at x = 0 or x = L due to the infinite potentials at the walls; hence, the wavefunction must be equal to zero at these points.

$$\psi(0) = A\sin(k0) + B\cos(k0) = B = 0 \quad (2\text{-}15a)$$

The first boundary condition reduces the wavefunction to $\psi(x) = A\sin(kx)$.
The next boundary condition at x = L now needs to be applied.

$$\psi(L) = A\sin(kL) = 0 \quad (2\text{-}15b)$$

There are two possible solutions to Equation 2-15b. The first solution is that A = 0; however, this would be a trivial solution since the wavefunction would equal to zero everywhere inside the box signifying that there is no particle. The other solution is that the sine is zero at x = L. The sine function is zero at 0, π, 2π, 3π, or some whole number multiple, n, of π. If the value of n is equal to zero, the wavefunction becomes zero everywhere in the box, which again would signify that there is no particle. As a result, the wavefunction for the problem becomes:

$$\psi(x) = A\sin\left(\frac{n\pi x}{L}\right) \quad (2\text{-}16)$$

where n =1, 2, 3, … and

$$k = \frac{n\pi}{L}. \tag{2-17}$$

The wavefunction now needs to be normalized which will determine the constant A. According to Equation 2-9, the square of the wavefunction (since the wavefunction here is real) must be integrated over all space which is from x = 0 to x = L and set equal to unity.

$$\int \psi^* \psi d\tau = \int_0^L \psi^* \psi d\tau = A^2 \int_0^L \sin^2\left(\frac{n\pi x}{L}\right) dx = 1$$

$$A^2\left[\frac{L}{2}\right] = 1$$

$$A = \sqrt{\frac{2}{L}}$$

The normalized wavefunction and the energy for the particle in a one-dimensional box are as follows (n = 1, 2, 3, ...).

$$\psi(x) = \sqrt{\frac{2}{L}} \sin\left(\frac{n\pi x}{L}\right) \tag{2-18}$$

$$E_n = \frac{\hbar^2 \pi^2 n^2}{2mL^2} = \frac{h^2 n^2}{8mL^2} \tag{2-19}$$

For a given system, the mass of the particle and the dimensions of the box are all a constant, k.

$$E_n = kn^2$$

Note that the energy difference between each energy level ($E_{n+1} - E_n$) increases with increasing value of n.

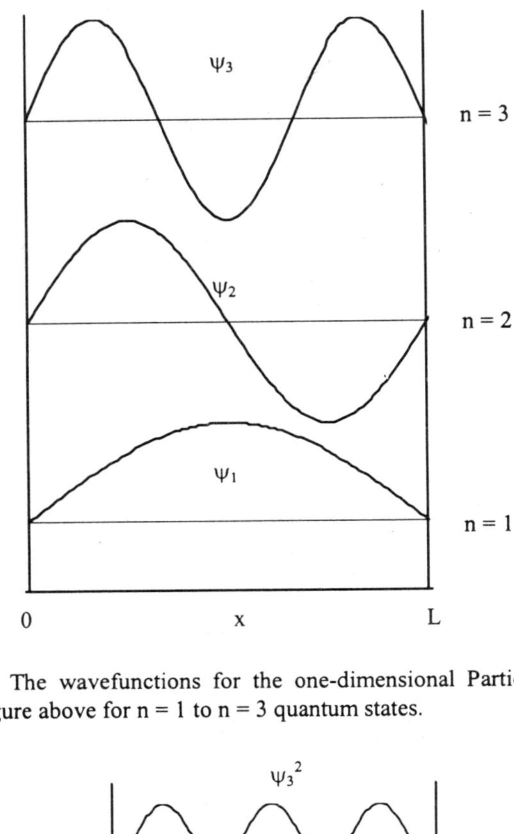

Figure 2-2. The wavefunctions for the one-dimensional Particle-in-a-Box are shown in the figure above for n = 1 to n = 3 quantum states.

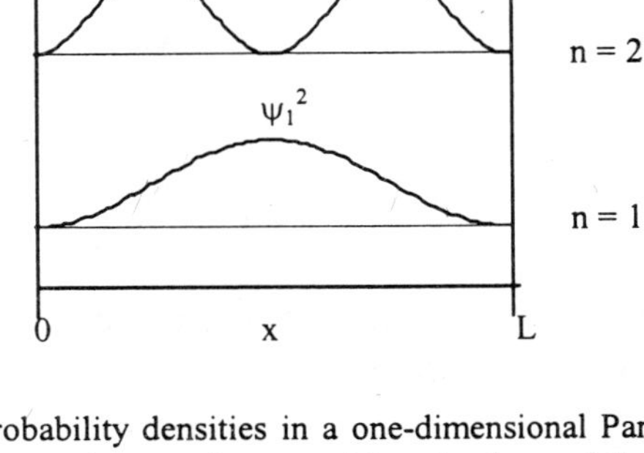

Figure 2-3. The probability densities in a one-dimensional Particle-in-a-Box system is shown for the n = 1 to n = 3 states. Note that beyond the ground-state, there are points with the box where the probability density is zero.

Note that quantization of the energy states for the particle has occurred due to the potential energy of the system. Only those states that will result in nodes in the wavefunction at the two walls of the box are allowed. At a node, the value of the wavefunction will become zero indicating that there is a zero probability of finding the particle at those points.

In Figure 2-2, the wavefunctions for the first several quantum states are shown. The probability of the particle at each point within the box for the first several states is shown in Figure 2-3. It is interesting to contrast the classical mechanical results with the quantum mechanical results that emerge from these figures. The classical result predicts an equal probability for the particle to occupy any point within the box. In addition, the classical result predicts any energy is possible with the ground-state energy (the lowest possible) as being zero. The quantum mechanical result demonstrates that the particle in the ground-state, n = 1, has its highest probability towards the middle of the box and the probability reaches a minimum as it approaches the infinite potential of the walls. In the n = 2 and higher states, note that nodes in the wavefunction form *within the box*. The particle probability at the nodal points of the wavefunction within the box are zero. This means that the particle has zero probability at these points within box even though the potential energy is still zero. This is only possible if the particle has wavelike properties. Also note that the degree of curvature of the wavefunction increases with increasing kinetic energy (increasing values of n). *The degree of curvature of the wavefunction is indicative of the amount of kinetic energy the particle possesses.*

Example 2-1

Problem: Find the probability of finding the particle in the first tenth (from x = 0 to x = L/10) of the box for n =1, 2, and 3 states.

Solution: The wavefunction is given by Equation (2-18).

$$\psi_n(x) = \sqrt{\frac{2}{L}}\sin\left(\frac{n\pi x}{L}\right)$$

The following integral will be used:

$$\int \sin^2(cx)dx = \frac{x}{2} - \left(\frac{1}{4c}\right)\sin(2cx)$$

To find the probability in a region, the probability density must be integrated over that region of space.

$$P_n = \int_0^{L/10} \left[\sqrt{\frac{2}{L}}\sin\left(\frac{n\pi x}{L}\right)\right]\left[\sqrt{\frac{2}{L}}\sin\left(\frac{n\pi x}{L}\right)\right]dx = \frac{2}{L}\int_0^{L/10} \sin^2\left(\frac{n\pi x}{L}\right)dx$$

$$P_n = \frac{2}{L}\left[\frac{L}{20} - \left(\frac{L}{4n\pi}\right)\sin\left(\frac{2n\pi}{10}\right)\right] = \left[\frac{1}{10} - \left(\frac{1}{2n\pi}\right)\sin\left(\frac{n\pi}{5}\right)\right]$$

The probability for each level in this region of the box can be computed by substitution of n.

$$\text{For n} = 1:\ P_1 = \frac{1}{10} - \frac{1}{2\pi}\sin\left(\frac{\pi}{5}\right) \cong 0.0064$$

$$\text{For n} = 2:\ P_2 = \frac{1}{10} - \frac{1}{4\pi}\sin\left(\frac{2\pi}{5}\right) \cong 0.024$$

$$\text{For n} = 3:\ P_3 = \frac{1}{10} - \frac{1}{6\pi}\sin\left(\frac{3\pi}{5}\right) \cong 0.050$$

The classical prediction is 0.1 for this region of the box for any energy of the particle. The quantum mechanical probability is much lower. The particle tends to "avoid" the walls where the potential is infinite. Also note that as the value of n approaches infinity, the classical result of 0.1 is obtained.

2.5 HERMITIAN OPERATORS

Hermitian operators are very important in quantum mechanics because their eigenvalues are real. As a result, hermitian operators are used to represent observables since an observation must result in a real number. Examples of hermitian operators include position, momentum, and kinetic and potential energy. An operator is hermitian if it satisfies the following relation:

$$\int \psi_m^* \hat{O} \psi_n d\tau = \left[\int \psi_n^* \hat{O} \psi_m d\tau\right]^* = \int \left(\hat{O} \psi_m\right)^* \psi_n d\tau \qquad (2\text{-}20)$$

for any two wavefunctions ψ_n and ψ_m. The term on the right of Equation 2-20 means take the complex conjugate of the operator and the wavefunction then multiply by the wavefunction ψ_n and integrate overall space. This definition ensures that eigenvalues of hermitian operators (i.e. observables) are real.

2.6 OPERATORS AND EXPECTATION VALUES

As defined in Section 2.5, any hermitian operator, $\hat{O}$, signifies a mathematical operation to be done on a wavefunction, ψ, which will yield a constant, o, if the wavefunction is an eigenfunction of the operator.

$$\hat{O}\psi = o\psi \qquad (2\text{-}21)$$

Next the complex conjugate of the wavefunction, ψ*, is multiplied to both sides of Equation 2-21 and integrated over all space.

$$\int \psi^* \hat{O} \psi d\tau = \int \psi^* o \psi d\tau = o \int \psi^* \psi d\tau$$

If the wavefunction is normalized, then the integral $\int \psi^* \psi d\tau$ is equal to one as shown in Equation 2-9. This leads directly to the value of the constant o.

$$< o >= \int \psi^* \hat{O} \psi d\tau \qquad (2\text{-}22)$$

As mentioned previously, the constant o corresponds to some physically observable quantity such as position, momentum, kinetic energy, or total energy of the system, and it is called the ***expectation value***. Since the expression in Equation 2-22 is being integrated over all space, the value obtained for the physically observable quantity corresponds to the *average* value of that quantity. This leads to the fourth postulate of quantum mechanics.

Postulate 4: *If the system is described by the wavefunction* ψ, *the mean value of the observable o is equal to the expectation value of the corresponding hermitian operator,* $\hat{O}$.
Postulate 4 leads to the generalized expression below that can be reduced to Equation 2-22 if the wavefunction is normalized.

$$<o>=\frac{\int \psi^* \hat{O} \psi d\tau}{\int \psi^* \psi d\tau} \tag{2-23}$$

Example 2-2
Problem: Determine the average position, <x> for the Particle-in-a-Box model problem for any state n.

Solution: The integral, which must be solved, is that given in Equation 2-22.

$$<x>=\int \psi^* \hat{x} \psi d\tau = \frac{2}{L}\int_0^L x \sin^2\left(\frac{n\pi x}{L}\right)dx = \frac{2}{L}\left(\frac{L^2}{4}\right) = \frac{L}{2}$$

This states that the average position of the particle is at the center of the box as is predicted by classical mechanics.

The types of integrals in Equation 2-23 are encountered often in quantum mechanics. Paul Dirac developed shorthand to represent these types of integrals called "bra-ket" notation. The integral in the numerator of Equation 2-23 is represented in "bra-ket" notation as follows:

$$\int \psi^* \hat{O} \psi d\tau \equiv \langle \psi | \hat{O} | \psi \rangle .$$

The "bra" of the wavefunction, $\langle \psi |$, represents the complex conjugate of the wavefunction. The "ket" of the wavefunction, $| \psi \rangle$, corresponds to the wavefunction which is operated on by the operator, $\hat{O}$. When the "bra" and "ket" are put together, it indicates that the product is to be integrated over all space. Equivalently, the integral in the denominator of Equation 2-23 is represented in "bra-ket" notation as follows:

$$\int \psi^* \psi d\tau = \langle \psi | \psi \rangle .$$

The value of this integral is unity if the wavefunctions are normalized and ψ and ψ* correspond to the same state. If ψ and ψ* correspond to different states, the value of the integral will be zero, and the wavefunctions are said to be ***orthogonal***. Wavefunctions that are orthogonal and normalized are called ***orthonormal***.

$$\langle \psi_m | \psi_n \rangle = \int \psi_m^* \psi_n d\tau = 1 \quad \text{(If n = m)}$$

$$= 0 \quad (\text{If n} \neq \text{m}) \tag{2-24}$$

2.7 THE HEISENBERG UNCERTAINTY PRINCIPLE

An interesting point to note in the Particle-in-a-Box model problem is that the ground-state energy is not zero as would be predicted by classical mechanics. The physical reason for this paradox has to do with uncertainties in knowing both the position and the momentum of the particle simultaneously due to the wavelike properties of the particle.

There is inherent error in any type of measurement. The standard deviation is an average range of measurements in a series of trials. As an example, suppose the following values were obtained for some measurement: 6.3, 6.8, 6.5, 6.2, and 6.9. The average value is 6.5. The individual trials deviate from this average by –0.2, 0.3, 0.0, -0.3, and 0.4. Simply taking an average of these deviations will result in some cancellations since some of the deviations are positive whereas others are negative. To avoid these cancellations, a root mean square (rms) ***uncertainty*** is determined whereby the square root of the mean of the square of each deviation is obtained. For this set of measurements, the uncertainty is 0.3. Four of the five measurements are within the range of 6.5 ± 0.3.

The analogous approach can be done in quantum mechanical systems. The square of the difference of the operator for an observable, $\hat{O}$, from the average, $<o>$, is taken: $(\hat{O}-<o>)^2$. The uncertainty squared, $(\Delta O)^2$, is the expectation value of this operator.

$$(\Delta O)^2 = \int \psi^* \left(\hat{O} - <o>\right)^2 \psi d\tau$$

This expression can be simplified by expansion. The <o> corresponds to a constant which can factored out of the integration. Assume the wavefunction is normalized. Then

$$\int \psi^* \left(\hat{O} - <o>\right)^2 \psi d\tau = \int \psi^* \left(\hat{O}^2 - 2\hat{O} <o> + <o>^2\right)\psi d\tau$$

$$= \int \psi^* \hat{O}^2 \psi d\tau - 2 <o> \int \psi^* \hat{O} \psi d\tau + <o>^2 \int \psi^* \psi d\tau$$

$$= <o^2> - 2<o><o> + <o>^2 = <o^2> - <o>^2.$$

The uncertainty, ΔO, is the square root of the expression above.

$$\Delta O = \sqrt{<o^2> - <o>^2} \qquad (2\text{-}25)$$

Example 2-3.

Problem: Determine the uncertainty in the momentum, Δp, for the ground-state energy of the Particle-in-a-Box model problem.

Solution: According to Equation 2-25, the following must be solved.

$$\Delta p = \sqrt{<p^2> - <p>^2}$$

The expression above states that the average of the square of the momentum, $<p^2>$, along with the square of the average momentum, $<p>^2$, must be determined. The average of the square of the momentum is determined as follows.

$$< p^2 >= \int \psi^* \hat{p}^2 \psi d\tau$$

$$= \frac{2}{L}\int_0^L \sin\left(\frac{\pi x}{L}\right)\left[-i\hbar\frac{d}{dx}\right]^2 \sin\left(\frac{\pi x}{L}\right)dx = \frac{2}{L}\left(-\hbar^2\right)\int_0^L \sin\left(\frac{\pi x}{L}\right)\left[\frac{d^2}{dx^2}\right]\sin\left(\frac{\pi x}{L}\right)dx$$

$$= -\hbar^2\left(\frac{-\pi^2}{L^2}\right)\left(\frac{2}{L}\right)\int_0^L \sin\left(\frac{\pi x}{L}\right)\sin\left(\frac{\pi x}{L}\right)dx = \left(\frac{h^2}{4\pi^2}\right)\left(\frac{\pi^2}{L^2}\right)(1) = \frac{h^2}{4L^2}$$

The average of the momentum squared is determined as follows.

$$< p >^2 = (\int \psi^* \hat{p} \psi d\tau)^2$$

$$= \left[\frac{2}{L}\int_0^L \sin\left(\frac{\pi x}{L}\right)\left[-i\hbar\frac{d}{dx}\right]\sin\left(\frac{\pi x}{L}\right)dx\right]^2$$

$$= \left[-\hbar^2\left(\frac{-\pi^2}{L^2}\right)\frac{2}{L}\int_0^L \sin\left(\frac{\pi x}{L}\right)\cos\left(\frac{\pi x}{L}\right)dx\right]^2 = \left[-\hbar^2\left(\frac{-\pi^2}{L^2}\right)(0)\right]^2 = 0$$

Note that the average momentum is zero as would be expected: the particle must have an equal average momentum towards each side of the box. The standard root mean square deviation in the momentum can now be solved.

$$\Delta p = \sqrt{< p^2 > - < p >^2} = \sqrt{\left(\frac{h^2}{4L^2}\right) - 0} = \frac{h}{2L}$$

The Heisenberg Uncertainty Principle states that for any system there are lower limits to the uncertainty of a given measurable observable. The product of the uncertainty of two conjugate measurable observables (see Section 1.2) is on the order of ½$\hbar$ or greater. Momentum and position along the same coordinate are examples of corresponding measurable observables.

$$\Delta q \Delta p \geq \frac{\hbar}{2} \tag{2-26}$$

The value of $\hbar$ is quite small and can be considered negligible in the macroscopic world. The Heisenberg Uncertainty Principle implies that the smaller the uncertainty in one observable, the greater the uncertainty in the other corresponding observable.

The Heisenberg Uncertainty Principle can now be readily applied to the ground-state of the Particle-in-a-Box. The uncertainty in the momentum of the particle has been previously determined in Example 2-3. The uncertainty in the position, Δx, for the ground-state is determined in a similar way as the momentum in Example 2-3.

For the ground-state of a Particle-in-a-Box:

$$\Delta p = \frac{h}{2L}$$

$$\Delta x = \sqrt{<x^2> - <x>^2} = \sqrt{\left(\frac{L^2}{3} - \frac{L^2}{2\pi^2}\right) - \left(\frac{L}{2}\right)^2} \cong 0.180L$$

$$\Delta p \Delta x \cong \left(\frac{h}{2L}\right)(0.180L) \cong 0.090h = 0.565\hbar > \frac{\hbar}{2}$$

As can be seen, the Heisenberg Uncertainty Principle is obeyed for the Particle-in-a-Box. As the box is made smaller (L is made increasingly smaller), the uncertainty in the position decreases whereas the uncertainty in the momentum increases. At the limit that L approaches zero, the uncertainty in the position becomes zero (the position of the particle would be confined to a point); however, the uncertainty in the momentum (and the energy) would become undefined indicating that there would be no knowledge of the particle's momentum. An understanding of why the ground-state energy for the Particle-in-a-Box is non-zero can now be rationalized. If the kinetic energy, and as a result the momentum, of the particle were zero, the position of the particle would also be precisely known because it is not in motion. This would be in direct violation of the Heisenberg Uncertainty Principle.

Point of Further Understanding
Consider a particle in a 1-dimensional box. What would happen to the quantization of the particle at the limit where L approaches infinity? The ground-state energy is equal to zero at this limit. Explain why this does not violate the Heisenberg Uncertainty Principle based upon the uncertainties in position and momentum.

2.8 PARTICLE IN A THREE-DIMENSIONAL BOX AND DEGENERACY

Particles normally are capable of travelling in three-dimensions, and the particle in a one-dimensional box can be readily expanded into three. The length of the box in each direction will be taken as L_x, L_y, and L_z in the x, y, and z coordinates respectively. The potential within the box is zero and it is infinite at the walls: 0, L_x, L_y, and L_z. The Hamiltonian for the particle will correspond to kinetic energy in the x, y, and z coordinates. The Schroedinger equation for the particle is as follows:

$$\hat{H}(x,y,z)\psi(x,y,z) = E\psi(x,y,z)$$

or

$$\frac{-\hbar^2}{2m}\left(\frac{\partial^2}{\partial x^2}+\frac{\partial^2}{\partial y^2}+\frac{\partial^2}{\partial z^2}\right)\psi(x,y,z) = E\psi(x,y,z)\,. \qquad (2\text{-}27)$$

The motion of the particle is independent in each dimension making this problem separable. Because the problem is separable, the Hamiltonian can be written as a sum of the Hamiltonians for each dimension.

$$\hat{H}(x, y, z) = \hat{H}(x) + \hat{H}(y) + \hat{H}(z) \tag{2-28}$$

In order to satisfy the Schroedinger equation (Equation 2-27), the wavefunction $\psi(x,y,z)$ must be the ***product*** of wavefunctions in each coordinate.

$$\psi(x, y, z) = \psi(x)\psi(y)\psi(z) \tag{2-29a}$$

In all separable systems, the Hamiltonian is represented as a sum along each independent variable, and the wavefunction for the system will be a product of the wavefunctions for each independent variable.

The system in each dimension is identical to a Particle-in-a-one-dimensional-Box. As a result, Equation 2-29a is a product of wavefunctions for the one-dimensional Particle-in-a-Box.

$$\psi(x, y, z) = \left(\sqrt{\frac{2}{L_x}} \sin\left(\frac{n_x \pi x}{L_x}\right)\right)\left(\sqrt{\frac{2}{L_y}} \sin\left(\frac{n_y \pi y}{L_y}\right)\right)\left(\sqrt{\frac{2}{L_z}} \sin\left(\frac{n_z \pi z}{L_z}\right)\right) \tag{2-29b}$$

The energy eigenvalues are found by applying Equation 2-29b to the Hamiltonian in Equation 2-27. The energy eigenvalues will depend on three different quantum numbers: n_x, n_y, and n_z corresponding to each coordinate.

$$E_{n_x, n_y, n_y} = \frac{h^2}{8m}\left(\frac{n_x^{\ 2}}{L_x^{\ 2}} + \frac{n_y^{\ 2}}{L_y^{\ 2}} + \frac{n_z^{\ 2}}{L_z^{\ 2}}\right) \tag{2-30}$$

where: $n_x = 1, 2, 3, \ldots$; $n_y = 1, 2, 3, \ldots$; $n_z = 1, 2, 3, \ldots$

Degeneracy occurs when there is more than one possible state with the same energy. Consider a particle in a three-dimensional box where the lengths in the x and y direction are equal: $L_x = L_y \neq L_z$. Under these circumstances, it is possible to have more than one state (n_x, n_y, n_z) with the same energy. As an example, the state 1, 3, 2 will have the same energy as 3, 1, 2. Degeneracy is an important concept as it occurs often in chemical systems.

Chemical Connection

The particle in a three-dimensional box can be used to model the translational energy of a gas phase atom or molecule. Consider a gas phase argon atom in a 1.00 m^3 square box at 300. K. The average thermal energy <E> of the argon atom is found by multiplying the temperature by Boltzmann's constant, k.

$$< E >= kT = 1.38x10^{-23} J \cdot K^{-1}(300.K) = 4.14x10^{-21} J$$

An argon atom has only three degrees of freedom: translational motion in the x, y, and z coordinates. If the translational energy of the argon atom is 4.14 x 10^{-21}J, what is a possible translational quantum state for the atom? How much energy is required to promote the atom one quantum level in each dimension? The energy differences between each translational quantum level is so small at these levels (i.e. a continuum) that in statistical thermodynamics it is approximated as infinitesimal.

PROBLEMS AND EXERCISES

2.1) Calculate the de Broglie wavelength for a supersonic jet aircraft with mass of 2.62×10^4 kg travelling at 2.55×10^3 km/hr. Is this wavelength significant relative to the size of a typical fighter aircraft? Now calculate the de Broglie wavelength for an electron (mass equal to 9.11×10^{-31} kg) with 13.7 eV of kinetic energy. (The electron volt, eV, is a convenient unit for describing the energy of small particle such as an electron: 1 eV = 1.6×10^{-19} J.)

2.2) For a Particle in a 1-dimensional Box, determine whether the following operators are Hermitian: a) position; b) momentum; and c) kinetic energy.

2.3) Demonstrate that the eigenfunctions for the Particle in a 1-dimensional Box are orthonormal.

2.4) For the particle in a 1-dimensional Particle-in-a-Box, determine the probability of the particle in the center $1/10^{th}$ of the box for the n = 1 and n = 2 states. Justify your results based on the shapes of the wavefunctions in this region as shown in Figure 2-2. How does this compare to what would be predicted classically?

2.5) For the Particle in a 1-dimensional Box, what is the probability that the particle is at x=L/2 in the n=1 and n=2 states?

2.6) Determine the energy eigenvalues and eigenfunctions for a particle free to travel from $x = -\infty$ to $x = \infty$. What occurs to the quantization of the particle's energy? What might you infer about the curvature of the eigenfunctions?

2.7) Calculate explicitly the momentum expectation value for the 1-dimensional Particle-in-a-Box for the n = 2 state.

2.8) Calculate the uncertainty in the momentum and position for a 1-dimensional Particle-in-a-Box for the n =10 and n =100 states. Is the product of the uncertainties within Heisenberg Uncertainty Principle limits?

2.9) Determine the uncertainty in the momentum and position for each dimension for a 3-dimensional Particle-in-a-Box in the ground-state. Why is it not possible for the energy to be zero in any one dimension even if it is not in the other dimensions?

2.10) Consider a particle in a 3-dimensional box with the following dimensions: $L_x = 2L_y = 2L_z$. Does this system have any degenerate states? Justify your answer.

Chapter 3

Rotational Motion

Rotational motion is an important topic in chemical systems as it will be used, in the chapters to follow, to describe the rotational motion of gas phase molecules and electronic motion in atoms and molecules. The model problems presented in this chapter will be the basis for modeling rotational motion throughout the remainder of the text.

3.1 PARTICLE-ON-A-RING

Consider a particle of mass m confined to a circle with a constant radius r as shown in Figure 3.1. The potential energy anywhere on the circle is defined as zero. The Hamiltonian, $\hat{H}$, for the particle in Cartesian coordinates is given below.

$$\hat{H}(x, y) = -\frac{\hbar^2}{2m}\left(\frac{\partial^2}{\partial x^2} + \frac{\partial^2}{\partial y^2}\right) \qquad (3\text{-}1)$$

The motion of the particle is not separable between the x and y-axes. The problem can be made separable by transforming the coordinates from Cartesian to polar coordinates. In polar coordinates, the variables become the radius of gyration, r, and the angle, ϕ, of the particle from the origin.

$$x = r\cos\phi \qquad\qquad y = r\sin\phi$$

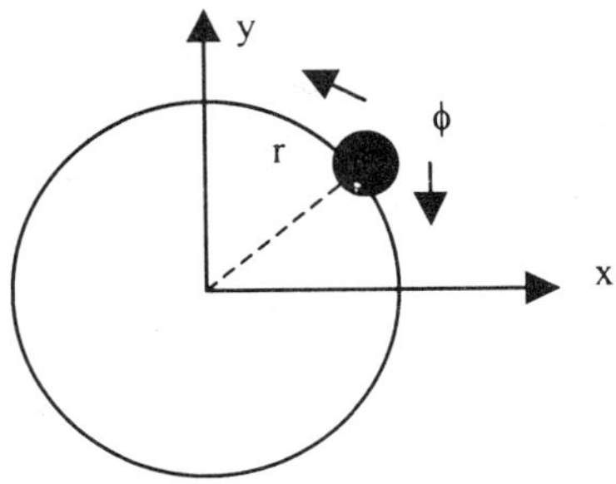

Figure 3-1. A particle confined to travel on a ring with a constant radius r.

The second derivatives with respect to x and y in Equation 3-1 are transformed into polar coordinates using the chain rule.

$$\frac{\partial^2}{\partial x^2}+\frac{\partial^2}{\partial y^2}=\frac{\partial^2}{\partial r^2}+\frac{1}{r}\frac{\partial}{\partial r}+\frac{1}{r^2}\frac{\partial^2}{\partial \phi^2}$$

The Hamiltonian in Equation 3-1 can now be written in terms of r and ϕ.

$$\hat{H}(r,\phi)=-\frac{\hbar^2}{2m}\left(\frac{\partial^2}{\partial r^2}+\frac{1}{r}\frac{\partial}{\partial r}+\frac{1}{r^2}\frac{\partial^2}{\partial \phi^2}\right)$$

Since the radius r is constant in this problem, all of the terms that involve derivatives with respect to r will be zero, reducing the Hamiltonian to just one variable, ϕ.

$$\hat{H}(\phi)=-\frac{\hbar^2}{2m}\left(\frac{1}{r^2}\frac{d^2}{d\phi^2}\right)=-\frac{\hbar^2}{2I}\frac{d^2}{d\phi^2} \qquad (3\text{-}2)$$

The moment of inertia, I, of the particle, is equal to its mass times the square of the radius of gyration r. The Hamiltonian in Equation 3-2 is very similar to the Hamiltonian for the one-dimensional Particle-in-a-Box problem (see

Section 2.4). This will result in the same functional forms for the wavefunctions in terms of the variable ϕ. The following wavefunction will be used:

$$\psi(\phi) = Ae^{ik\phi} + Be^{-ik\phi}. \tag{3-3}$$

The constants A and B are normalization constants, and the constant k will be determined by the boundary conditions for the particle. The Schroedinger equation and the energy eigenvalue becomes:

$$\hat{H}(\phi)\psi(\phi) = -\frac{\hbar^2}{2I}\frac{d^2}{d\phi^2}\left(Ae^{ik\phi} + Be^{-ik\phi}\right) = E\psi(\phi)$$

$$E = \frac{k^2\hbar^2}{2I}. \tag{3-4}$$

The boundary conditions must now be applied to the system. There are no points along the circular path where the potential becomes infinite as in the case of the Particle-in-a-Box; hence, the wavefunction does not need to truncate at any points. The wavefunction, however, must be continuous and single valued at a given point (see Section 2.3). This requires that the value of the wavefunction must be the same at $\phi = 0$ and 2π.

$$\psi(0) = \psi(0 + 2\pi)$$

$$Ae^{ik0} + Be^{-ik0} = Ae^{ik0}e^{ik2\pi} + Be^{-ik0}e^{-ik2\pi}$$

$$A + B = Ae^{ik2\pi} + Be^{-ik2\pi} \tag{3-5}$$

The equality in Equation 3-5 is satisfied only if k is an integer (since then $e^{ik2\pi}$ and $e^{-ik2\pi}$ individually are equal to 1). The symbol for k is traditionally m_l known as the magnetic quantum number when referring to electronic states. The possible values for m_l are as follows:

$$m_l = 0, \pm 1, \pm 2, \pm 3, \ldots$$

The allowed energy values become:

$$E = \frac{m_l^{\,2}\hbar^2}{2I} \quad \text{where } m_l = 0, \pm 1, \pm 2, \pm 3, \ldots \tag{3-6}$$

The final task is to normalize the wavefunction. Since the value of m_l can be both positive and negative integers, the wavefunction can be reduced to just one exponential term.

$$\psi(\phi) = Ae^{im_l\phi}$$

The limits of integration for normalization will be from 0 to 2π since this covers the entire circular path.

$$\int \psi^* \psi d\tau = A^2 \int_0^{2\pi} \left(e^{-im_l\phi}\right)\left(e^{im_l\phi}\right) d\phi = 1$$

$$A = \sqrt{\frac{1}{2\pi}}$$

The normalized wavefunction for the Particle-on-a-Ring becomes:

$$\psi(\phi) = \sqrt{\frac{1}{2\pi}} e^{im_l\phi}; \qquad m_l = 0, \pm 1, \pm 2, \ldots \tag{3-7}$$

A physical connection needs to be made of the sign on the m_l quantum states. Positive and negative signs indicate direction, and in this case, the direction must be the direction of rotation: clockwise or counterclockwise. To confirm this, the angular momentum, L, of the particle can be determined. Since the rotation is confined to the x-y plane, the only non-zero component of the angular momentum of the particle is along the z-axis, L_z.

$$L_z = xp_y - yp_x$$

The angular momentum operator along the z-axis in polar coordinates is given as:

$$\hat{L}_z = \frac{\hbar}{i}\left(\frac{d}{d\phi}\right), \tag{3-8}$$

The angular momentum expectation value, $< L_z >$, is determined as follows:

$$< L_z > = \langle \psi(\phi) | \hat{L}_z | \psi(\phi) \rangle$$
$$= \int_0^{2\pi} \left[\left(\sqrt{\frac{1}{2\pi}} e^{-im_l\phi} \right) \frac{\hbar}{i} \left(\frac{d}{d\phi} \right) \left(\sqrt{\frac{1}{2\pi}} e^{im_l\phi} \right) \right] d\phi = m_l \hbar. \tag{3-9}$$

Based on the left-hand rule, positive angular momentum (for positive values of m_l) indicates a clockwise rotation whereas negative values indicate a counterclockwise rotation.

It is interesting to note that in this system, the ground-state energy ($m_l =$ 0) is zero unlike in the Particle-in-a-Box system. There are no points along the circular path that require the wavefunction to become zero; hence, it is possible for the wavefunction to have no curvature. As discussed previously in the Particle-in-a-Box system (see Section 2.4), the degree of curvature in a wavefunction is related to the amount of kinetic energy that the particle possesses. Since the ground-state wavefunction is a constant, there is no curvature to the wavefunction and the particle has no kinetic energy or correspondingly angular momentum. The probability density of the particle in the ground-state is the same throughout the entire circular path since the wavefunction is a standing wave of constant amplitude. The real portion of the wavefunctions for the m_l states of +1 and +2 are shown in Figures 3-2a and 3-2b. The m_l states of –1 and –2 are same as shown Figure 3-2 except the waves are now inverted. In these higher states, curvature in the wavefunctions emerges indicating a non-zero kinetic energy. The anti-nodes of the wavefunction lie above and below the ring.

(a) $m_l = +1$

(b) $m_l = +2$

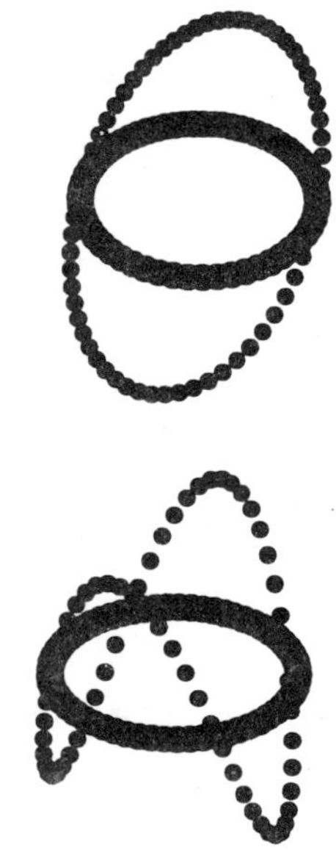

Figure 3-2. The real part of the first and second excited state wavefunctions for a particle confined to a ring of constant radius are shown.

3.2 PARTICLE-ON-A-SPHERE

This model problem will be used to describe the rotational motion of molecules and the electron motion in atoms. Consider a particle of mass m free to rotate on the surface of a sphere with a constant radius r. The potential on the sphere is zero; hence, the Hamiltonian will have only the kinetic energy operator for each coordinate.

$$\hat{H}(x,y,z) = -\frac{\hbar^2}{2m}\left(\frac{\partial^2}{\partial x^2}+\frac{\partial^2}{\partial y^2}+\frac{\partial^2}{\partial z^2}\right) = -\frac{\hbar^2}{2m}\nabla^2$$

The second derivative over each coordinate is called the del-squared operator: ∇^2. The motion of the particle is not separable in Cartesian coordinates. A coordinate transformation to spherical coordinates (r, θ, φ) assures separability.

$$x = r\sin\theta\cos\phi \qquad y = r\sin\theta\sin\phi \qquad z = r\cos\theta \tag{3-10}$$

The del-squared operator must now be converted to spherical coordinates.

$$\nabla^2 = \frac{1}{r}\left(\frac{\partial^2}{\partial r^2}\right)r + \frac{1}{r^2}\Lambda^2 \tag{3-11}$$

The Λ^2 contains the operations with the angular variables θ and φ and is called the ***legendrian***.

$$\Lambda^2 = \frac{1}{\sin^2\theta}\left(\frac{\partial^2}{\partial\phi^2}\right) + \frac{1}{\sin\theta}\left(\frac{\partial}{\partial\theta}\right)\sin\theta\left(\frac{\partial}{\partial\theta}\right) \tag{3-12}$$

The Hamiltonian for a Particle-on-a-Sphere can now be written in terms of spherical coordinates.

$$\hat{H}(r,\theta,\phi) = -\frac{\hbar^2}{2m}\left[\frac{1}{r}\left(\frac{\partial^2}{\partial r^2}\right)r + \frac{1}{r^2}\Lambda^2\right] \tag{3-13}$$

Since r is constant in this system, the second derivative with respect to r is zero reducing the Hamiltonian to just the legendrian part of ∇^2.

$$\hat{H}(\theta,\phi) = -\frac{\hbar^2}{2mr^2}\Lambda^2 = -\frac{\hbar^2}{2I}\Lambda^2$$
$$= -\frac{\hbar^2}{2I}\left[\frac{1}{\sin^2\theta}\left(\frac{\partial^2}{\partial\phi^2}\right) + \frac{1}{\sin\theta}\left(\frac{\partial}{\partial\theta}\right)\sin\theta\left(\frac{\partial}{\partial\theta}\right)\right] \tag{3-14}$$

The Schroedinger equation for the Particle-on-a-Sphere becomes:

$$-\left(\frac{\hbar^2}{2I}\right)\Lambda^2\psi(\theta,\phi) = E\psi(\theta,\phi),$$

$$-\frac{\hbar^2}{2I}\left[\frac{1}{\sin^2\theta}\left(\frac{\partial^2}{\partial\phi^2}\right)\psi(\theta,\phi)+\frac{1}{\sin\theta}\left(\frac{\partial}{\partial\theta}\right)\sin\theta\left(\frac{\partial}{\partial\theta}\right)\psi(\theta,\phi)\right] \quad (3\text{-}15)$$
$$= E\psi(\theta,\phi).$$

Since Equation 3-15 is a sum of derivatives with respect to θ and ϕ, it is separable. As a result, the wavefunction ψ(θ,ϕ) will be a product of wavefunctions in terms of θ and ϕ as previously discussed in Section 2.7.

$$\psi(\theta,\phi) = \Theta(\theta)\Phi(\phi)$$

The first part of the Schroedinger equation in Equation 3-15 involves taking the second derivative of the wavefunction with respect to ϕ. This is identical to the operation as seen previously for the Particle-on-a-Ring; hence, the wavefunction Φ(ϕ) will be the same as in the Particle-on-a-Ring.

$$\left(\frac{\partial^2}{\partial\phi^2}\right)\Theta(\theta)\Phi(\phi) = -m_l^{\,2}\Theta(\theta)\Phi(\phi) \quad (3\text{-}16)$$

Substitution of Equation 3-16 into Equation 3-15 results in the following differential equation:

$$-\frac{\hbar^2}{2I}\left[\left(\frac{-m_l^{\,2}}{\sin^2\theta}\right)\Theta(\theta)\Phi(\phi)+\frac{1}{\sin\theta}\left(\frac{\partial}{\partial\theta}\right)\sin\theta\left(\frac{\partial}{\partial\theta}\right)\Theta(\theta)\Phi(\phi)\right] \quad (3\text{-}17)$$
$$= E\Theta(\theta)\Phi(\phi).$$

The solution to this differential equation is well known. The wavefunctions Θ(θ) that satisfy Equation 3-17 are the ***associated Legendre polynomials***. The associated Legendre polynomials, $P_l^{|m_l|}(z)$, for a variable z are obtained from the following recursion relationship:

$$P_l^0(z) = \frac{1}{2^l l!}\frac{d^l}{dz^l}\left[\left(z^2 - 1\right)^l\right],$$

$$P_l^{|m_l|}(z) = (1 - z^2)^{m_l/2}\frac{d^{m_l}}{dz^{m_l}}P_l^0(z).$$

The variable for the function Θ is actually cos θ; hence, the associated Legendre polynomials are generated in terms of z, and then z is replaced by cos θ throughout. The normalization constant is obtained by integrating from θ = 0 to π.

$$\sqrt{\frac{(2l+1)(l-|m_l|)!}{2(l+|m_l|)!}}$$

The normalized associated Legendre polynomials, Θ(θ) are given in the following expression, and the first several are listed in Table 3-1.

$$\Theta(\theta) = \sqrt{\frac{(2l+1)(l-|m_l|)!}{2(l+|m_l|)!}}P_l^{|m_l|}(\cos\theta) \qquad (3\text{-}18)$$

The product of the normalized associated Legendre polynomials along with the Particle-on-a-Ring functions are known as the ***spherical harmonics*** symbolized as Y_{lm_l}.

$$Y_{lm_l} = \Theta(\theta)\Phi(\phi)(-1)^{[m_l+|m_l|]/2} \qquad (3\text{-}19)$$

The arbitrary phase factor

$$(-1)^{[m_l+|m_l|]/2}$$

is introduced to conform to common conventions and its value is +1 or −1. Table 3-1 lists the first few spherical harmonic wavefunctions. When the legendrian, Λ^2, is applied to a spherical harmonic wavefunction, the following eigenvalue results:

$$\Lambda^2 Y_{lm_l} = -l(l+1)Y_{lm_l}. \qquad (3\text{-}20)$$

The values of l and m_l are integers such that $l = 0,1,2,3,...$ and $m_l = l, l-1,...,-l$. Though the determination of the particle's state requires specification of both l and m_l, the energy eigenvalue for the Particle-on-a-Sphere is dependent only on the l quantum number.

$$E = \frac{\hbar^2}{2I} l(l+1); \qquad \text{where } l = 0, 1, 2, 3, \ldots \qquad (3\text{-}21)$$

For each l quantum level, there are $2l + 1$ values of m_l resulting in $2l + 1$ degenerate states.

Table 3-1. The normalized associated Legendre polynomials, $\Theta_{lm_l}(\theta)$ and spherical harmonic wavefunctions, $Y_{lm_l}(\theta,\phi)$ up to $l = 3$

l	m_l	$\Theta_{lm_l}(\theta)$	$Y_{lm_l}(\theta,\phi)$
0	0	$\frac{\sqrt{2}}{2}$	$\frac{1}{2\sqrt{\pi}}$
1	0	$\sqrt{\frac{3}{2}}\cos\theta$	$\frac{1}{2}\sqrt{\frac{3}{\pi}}\cos\theta$
1	± 1	$\frac{\sqrt{3}}{2}\sin\theta$	$\mp\frac{1}{2}\sqrt{\frac{3}{2\pi}}(\sin\theta)e^{\pm i\phi}$
2	0	$\frac{\sqrt{10}}{4}(3\cos^2\theta - 1)$	$\frac{1}{4}\sqrt{\frac{5}{\pi}}(3\cos^2\theta - 1)$
2	± 1	$\frac{\sqrt{15}}{2}(\cos\theta\sin\theta)$	$\mp\frac{1}{2}\sqrt{\frac{15}{2\pi}}(\cos\theta\sin\theta)e^{\pm i\phi}$
2	± 2	$\frac{\sqrt{15}}{4}(\sin^2\theta)$	$\frac{1}{4}\sqrt{\frac{15}{2\pi}}(\sin^2\theta)e^{\pm 2i\phi}$
3	0	$\sqrt{\frac{7}{8}}(2 - 5\sin^2\theta)\cos\theta$	$\frac{1}{4}\sqrt{\frac{7}{\pi}}(2 - 5\sin^2\theta)\cos\theta$
3	± 1	$\sqrt{\frac{21}{32}}[(5\cos^2\theta - 1)\sin\theta]$	$\mp\frac{1}{8}\sqrt{\frac{21}{\pi}}[(5\cos^2\theta - 1)\sin\theta]e^{\pm i\phi}$
3	± 2	$\frac{\sqrt{105}}{4}\cos\theta\sin^2\theta$	$\frac{1}{4}\sqrt{\frac{105}{2\pi}}(\cos\theta\sin^2\theta)e^{\pm 2i\phi}$
3	± 3	$\frac{\sqrt{70}}{8}\sin^3\theta$	$\mp\sqrt{\frac{5}{\pi}}(\sin^3\theta)e^{\pm 3i\phi}$

Example 3-1

Problem: (a) Confirm that Y_{1-1} is an eigenfunction of the Particle-on-a-Sphere Hamiltonian, and (b) that it is normalized.

Solution: (a) The Y_{1-1} wavefunction is found in Table 3-1.

$$Y_{1-1} = \frac{1}{2}\left(\frac{3}{2\pi}\right)^{\frac{1}{2}}(\sin\theta)e^{-i\phi} = N(\sin\theta)e^{-i\phi}$$

N corresponds to the normalization constant. The Schroedinger equation becomes:

$$-\frac{\hbar^2}{2I}\Lambda^2 Y_{1-1} = EY_{1-1}$$

$$-\frac{\hbar^2}{2I}\left[\frac{1}{\sin^2\theta}\left(\frac{\partial^2}{\partial\phi^2}\right)N(\sin\theta)e^{-i\phi} + \frac{1}{\sin\theta}\left(\frac{\partial}{\partial\theta}\right)\sin\theta\left(\frac{\partial}{\partial\theta}\right)N(\sin\theta)e^{-i\phi}\right]$$
$$= EY_{1-1}$$

$$-\frac{\hbar^2}{2I}\left[\frac{-1}{\sin^2\theta}N(\sin\theta)e^{-i\phi} + \frac{1}{\sin\theta}\left(\frac{\partial}{\partial\theta}\right)(\sin\theta)N(\cos\theta)e^{-i\phi}\right] = EY_{1-1}$$

$$-\frac{\hbar^2}{2I}\left[-N\left(\sin^{-1}\theta\right)e^{-i\phi} + \frac{1}{\sin\theta}\left(\cos^2\theta - \sin^2\theta\right)Ne^{-i\phi}\right] = EY_{1-1}$$

$$-\frac{\hbar^2}{2I}\left[-N\left(\sin^{-1}\theta\right)e^{-i\phi} + \frac{1}{\sin\theta}\left(1 - 2\sin^2\theta\right)Ne^{-i\phi}\right] = EY_{1-1}$$

$$-\frac{\hbar^2}{2I}\left[-N\left(\sin^{-1}\theta\right)e^{-i\phi} + N\left(\sin^{-1}\theta\right)e^{-i\phi} - 2N(\sin\theta)e^{-i\phi}\right] = EY_{1-1}$$

$$\frac{\hbar^2}{I}Y_{1-1} = EY_{1-1}$$

$$E = \frac{\hbar^2}{I}$$

The energy calculated is the same result that is obtained directly from Equation 3-21.

$$E = \frac{\hbar^2}{2I} l(l+1) = \frac{\hbar^2}{2I} 1(1+1) = \frac{\hbar^2}{I}$$

(b) If Y_{1-1} is normalized; the following integral must be equal to one:

$$\langle Y_{1-1} | Y_{1-1} \rangle = \int Y_{1-1}^{*} Y_{1-1} d\tau = 1 .$$

The infinitesimal volume element dτ must be converted to spherical coordinates.

$$d\tau = dxdydz = r^2 dr \sin\theta d\theta d\phi$$

Since r is constant for the Particle-on-a-Sphere, dτ will be in terms of θ and ϕ only.

$$d\tau = \sin\theta d\theta d\phi$$

The integral becomes:

$$\int Y_{1-1}^{*} Y_{1-1} d\tau = \tfrac{1}{4}\left(\tfrac{3}{2\pi}\right) \int_{0}^{2\pi} \int_{0}^{\pi} [(\sin\theta) e^{i\phi}][(\sin\theta) e^{-i\phi}][\sin\theta d\theta d\phi]$$

$$\frac{1}{4}\left(\frac{3}{2\pi}\right)\left[\int_0^{\pi} \sin^3 d\theta\right]\left[\int_0^{2\pi} d\phi\right] = \frac{1}{4}\left(\frac{3}{2\pi}\right)\left(\frac{4}{3}\right)(2\pi) = 1$$

This confirms that the Y_{1-1} wavefunction is normalized.

The angular momentum for the particle can now be determined. When the particle is confined to rotate in only two-dimensions (i.e. confined to rotate on a ring), the angular momentum is parallel to the z-axis and is fully determined by the value of m_l. In three-dimensional rotation, the angular momentum need not be parallel to the z-axis and may also have components in the x and y-axes. The operators for the components of the angular momentum L in Cartesian coordinates are as follows:

$$\hat{L}_x = -i\hbar\left(y\frac{\partial}{\partial z} - z\frac{\partial}{\partial y}\right),$$

$$\hat{L}_y = -i\hbar\left(z\frac{\partial}{\partial x} - x\frac{\partial}{\partial z}\right),$$

$$\hat{L}_z = -i\hbar\left(x\frac{\partial}{\partial y} - y\frac{\partial}{\partial x}\right).$$

The component angular momentum operators can be transformed to spherical coordinates by using the chain-rule.

$$\hat{L}_x = -i\hbar\left(-\sin\phi\frac{\partial}{\partial\theta} - \frac{\cos\phi}{\tan\theta}\frac{\partial}{\partial\phi}\right) \quad (3\text{-}22)$$

$$\hat{L}_y = -i\hbar\left(\cos\phi\frac{\partial}{\partial\theta} - \frac{\sin\phi}{\tan\theta}\frac{\partial}{\partial\phi}\right) \quad (3\text{-}23)$$

$$\hat{L}_z = -i\hbar\frac{\partial}{\partial\phi} \quad (3\text{-}24)$$

The square of the angular momentum, L^2, can be found from the angular momentum component operators. The square of the angular momentum is a scalar quantity as it represents the dot product of $\vec{L}\cdot\vec{L}$.

$$\hat{L}^2 = \hat{L}_x{}^2 + \hat{L}_y{}^2 + \hat{L}_z{}^2$$

By combining with Equations 3-22, 3-23, and 3-24, the total angular momentum squared operator is obtained, and, to no surprise, it is proportional to the legendrian.

$$\hat{L}^2 = -\hbar^2\left(\frac{1}{\sin^2\theta}\frac{\partial^2}{\partial\phi^2} + \frac{1}{\sin\theta}\frac{\partial}{\partial\theta}\sin\theta\frac{\partial}{\partial\theta}\right) = -\hbar^2\Lambda^2 \qquad (3\text{-}25)$$

When the legendrian acts on the spherical harmonic wavefunctions, the result is as given in Equation 3-20.

$$\hat{L}^2 Y_{lm_l} = -\hbar^2\Lambda^2 Y_{lm_l} = \hbar^2 l(l+1) \qquad (3\text{-}26)$$

The magnitude of the angular momentum will be the square root of Equation 3-26.

$$\textit{magnitude of the angular momentum} = \hbar\sqrt{l(l+1)} \qquad (3\text{-}27)$$

As can be seen by Equation 3-27, the angular momentum is quantized.

The spherical harmonic wavefunctions are eigenfunctions only of the z angular momentum operator and the overall angular momentum squared operator.

$$\hat{L}_z Y_{lm_l} = \text{constant}\cdot Y_{lm_l}\,; \qquad \hat{L}^2 Y_{lm_l} = \text{constant}'\cdot Y_{lm_l}$$

The spherical harmonics are not eigenfunctions of the x and y angular momentum operators. This means that only the overall magnitude of angular momentum can be determined along with the magnitude in the z-coordinate. The magnitude in the z-coordinate is determined by applying the z angular momentum operator to the spherical harmonic wavefunctions.

$$\hat{L}_z Y_{lm_l} = m_l \hbar Y_{lm_l} \qquad (3\text{-}28)$$

As an example, for $l = 2$, the possible values of m_l are –2, -1, 0, 1, and 2. The magnitude of the angular momentum is $\hbar\sqrt{6}$ and the z-component is one of the possible five values: -2$\hbar$, -$\hbar$, 0, $\hbar$, and 2$\hbar$. Notice that in every case the z-component of the angular momentum is always less than the total angular momentum of the particle. This means that the angular momentum vector cannot lie parallel to the z-axis. Only for large values of l, such as for macroscopic objects (i.e. a ball), is the value of $\sqrt{l(l+1)}$ close to l such that the particle can be said to rotate *solely* about the z-axis.

Example 3-2

Problem: Consider the following two different masses undergoing rotational motion:

a) an electron (mass 9.11×10^{-31} kg)

b) a macroscopic particle (mass = 0.020 kg)

In both cases, the masses are rotating at 1.95 revolutions per second with a radius of gyration of 1.0 cm. Determine the l quantum number and the smallest angle θ that the angular momentum vector makes from the z-axis for each particle.

Solution: From classical mechanics, the magnitude of the angular momentum, L, is equal to the angular speed, ω, in radians per second times the moment of inertia, I. This can be related to the quantum mechanical expression given in Equation 3-27.

$$L = I\omega = \hbar\sqrt{l(l+1)}$$

The smallest angle θ that the angular momentum vector makes to the z-axis corresponds to the state that the z-component of the angular momentum is maximized. This occurs when $m_l = l$. The angle θ is determined by recognizing that cosine θ is equal to the magnitude of the z-component of the angular momentum divided by the magnitude of the angular momentum.

$$\cos\theta = \frac{L_z}{L} = \frac{m_l\hbar}{\hbar\sqrt{l(l+1)}} = \frac{m_l}{\sqrt{l(l+1)}}$$

The minimum angle, θ_{min}, occurs when $m_l = l$ as mentioned previously.

$$\cos\theta_{\min} = \frac{l}{\sqrt{l(l+1)}}$$

a) For the mass of an electron, the value of l can be determined as follows:

$$I = mr^2 = (9.11x10^{-31}\,kg)(1.0x10^{-2}\,m)^2 = 9.11x10^{-35}\,kg \cdot m^2$$

$$\sqrt{l(l+1)} = \frac{I\omega}{\hbar} = \frac{(9.11x10^{-35}\,kg \cdot m^2)(1.95 rev \cdot s^{-1})(2\pi \cdot rev^{-1})}{1.06x10^{-34}\,kg \cdot m^2 \cdot s^{-1}} = 10.5$$

$$l^2 + l - (10.5)^2 = 0$$

$$l = 10$$

The value of θ_{min} is determined as follows:

$$\cos\theta_{\min} = \frac{10}{10.5} \approx 0.95 \qquad \theta_{\min} \approx 18^\circ$$

b) For the case of the macroscopic particle (mass = 0.020 kg):

$$I = mr^2 = (0.020 kg)(1.0x10^{-2}\,m)^2 = 2.0x10^{-6}\,kg \cdot m^2$$

$$\sqrt{l(l+1)} = \frac{I\omega}{\hbar} = \frac{(2.0x10^{-6}\,kg \cdot m^2)(1.95 rev \cdot s^{-1})(2\pi \cdot rev^{-1})}{1.06x10^{-34}\,kg \cdot m^2 \cdot s^{-1}} = 2.31x10^{29}$$

$$l \cong \sqrt{2.31x10^{29}} = 4.81x10^{14}$$

$$\cos\theta_{\min} \cong \frac{4.81x10^{14}}{4.81x10^{14}} = 1 \qquad \theta_{\min} \cong 0^\circ$$

As can be seen in this example, the rotation of particles can be essentially confined to a plane only when the value of l is large such as in macroscopic particles.

PROBLEMS AND EXERCISES

3.1) Consider the rotation of the H atom in a HI molecule confined to rotate in a plane (a restriction that will be removed in a subsequent problem). Since the I atom is so much more massive than the H atom, it can be viewed as stationary. The radius of gyration will be taken as the bond length (approximately 160 pm). What wavelength of radiation is needed to undergo a transition from the ground-state to the first excited state if a) the hydrogen atom is ^{1}H and b) ^{2}H?

3.2) Confirm that the wavefunctions for the Particle-on-a-Ring are orthogonal.

3.3) Calculate a) the m_l energy level and b) the angular momentum for a wheel with a mass of 15.0 kg and a radius of 38.1 cm rotating in a plane at 45.0 rpm.

3.4) Confirm that Y_{10} and Y_{11} as given in Table 3-1 are a) eigenfunctions of the Particle-on-a-Sphere model problem; b) normalized; and c) orthogonal.

3.5) Repeat the calculation in Problem 3.1 by allowing the H atom to rotate freely in 3-dimensions. What wavelength of radiation is needed to undergo a transition from the ground-state to the first excited state for each type of H atom? What angle θ will the angular momentum vector make from the z-axis when the H atom is in the Y_{20} excited state?

3.6) Confirm that the Y_{21} is not an eigenfunction of the x or y angular momentum operators but is an eigenfunction of the z angular momentum and overall angular momentum squared operators.

Chapter 4

Techniques of Approximation

There are very few problems for which the Schroedinger equation can be solved for exactly, so methods of approximation are needed in order to tackle these problems. The two basic methods of approximation are *variation* and *perturbation* theories. In variation theory, an initial educated guess is made as to the shape of the wavefunction, which is then optimized to approximate the true wavefunction for the problem. In perturbation theory, the Schroedinger equation is separated into parts in which the solution is known (from previously solved problems or model problems) and parts that represent changes or "perturbations" from the known problem. The wavefunctions from the part of the Schroedinger equation in which the solution is known are used as a starting point and then modified to approximate the true wavefunction for the Schroedinger equation of interest. Both theories are important and powerful problem solving techniques that will be used throughout the rest of the text.

4.1 VARIATION THEORY

The first step is to write the Hamiltonian for the problem. Then an educated guess is made at a reasonable wavefunction called formally the trial wavefunction, ψ_{trial}. The trial wavefunction will have one or more adjustable parameters, p_i, that will be used for optimization. An energy expectation value in terms of the adjustable parameters, ε, is obtained by using the same form as in Equation 2-23.

$$\varepsilon = \frac{\langle \psi_{trial} | \hat{H} | \psi_{trial} \rangle}{\langle \psi_{trial} | \psi_{trial} \rangle} \tag{4-1}$$

The term in the denominator of Equation 4-1 is needed since the trial wavefunction is most likely not normalized.

Variation theory states that the energy expectation value ε is greater than or equal to the true ground-state energy, E_0, of the system. The equality occurs only when the trial wavefunction is the true ground-state wavefunction of the system.

$$\varepsilon \geq E_0 \qquad \text{(for any } \psi_{trial}\text{)} \tag{4-2}$$

Since ε is a function of the yet undetermined adjustable parameters p_i, the value of ε can be optimized by taking the derivative of ε with respect to each adjustable parameter and setting it equal to zero. A value for each parameter is then obtained for the optimized energy of the ground-state.

$$\frac{\partial \varepsilon}{\partial p_i} = 0 \tag{4-3}$$

Variation theory can be proven as follows. Take the trial wavefunction, ψ_{trial}, as a linear combination of the true eigenfunctions of the Hamiltonian, $\hat{H}$.

$$\psi_{trial} = \sum_n c_n \psi_n$$

Since ψ_n is an eigenfunction of the Hamiltonian of the system, applying $\hat{H}$ to ψ_n will result in an energy eigenvalue E_n.

$$\hat{H}\psi_n = E_n \psi_n$$

Now consider the following integral:

$$\int \psi^*_{trial} (\hat{H} - E_0) \psi_{trial} d\tau = \sum_n \sum_{n'} c^*_n c_{n'} \int \psi^*_n (\hat{H} - E_0) \psi_{n'} d\tau$$

$$= \sum_{n}\sum_{n'} c_n^* c_{n'} (E_{n'} - E_0) \int \psi_n^* \psi_{n'} d\tau \, .$$

Since the ψ_n eigenfunctions are orthonormal (see Equation 2-24), the previous integral is zero when $n \neq n'$ and one when $n = n'$.

$$\sum_{n}\sum_{n'} c_n^* c_{n'} (E_{n'} - E_0) \int \psi_n^* \psi_{n'} d\tau = \sum_{n} c_n^* c_n (E_n - E_0) \geq 0$$

The result above must be positive since

$$E_n \geq E_0 \text{ and } c_n^* c_n = |c_n|^2$$

are positive. Therefore:

$$\int \psi_{trial}^* (\hat{H} - E_0) \psi_{trial} d\tau = \int \psi_{trial}^* \hat{H} \psi_{trial} d\tau - E_0 = \varepsilon - E_0 \geq 0$$

or

$$\varepsilon \geq E_0$$

completing the proof.

Variation theory states that the energy calculated from any trial wavefunction will never be less than the true ground-state energy of the system. This means that the smaller the value of ε, the closer it is to the true ground-state energy of the system and the more ψ_{trial} represents the true ground-state wavefunction. The trial wavefunction is set up with one or more adjustable parameters, p_i, making the function flexible to minimize the value of ε. An n number of adjustable parameters will set up an n number of differential equations:

$$\left(\frac{\partial \varepsilon}{\partial p_1} = 0; \frac{\partial \varepsilon}{\partial p_2} = 0; \cdots; \frac{\partial \varepsilon}{\partial p_n} = 0 \right).$$

Increasing the number of adjustable parameters improves the result, however it also increases the complexity of the problem. The variational approach is demonstrated in the following example.

Example 4-1

Problem: A particle with a mass m is confined to a 1-dimensional box. Use the following trial wavefunction:

$$\psi_{trial} = N\left\{\left(x - \frac{x^2}{L}\right) + p\left(x - \frac{x^2}{L}\right)^2\right\}$$

where N is the normalization constant and p is the adjustable parameter. Note that this function is well behaved at the boundary conditions since the function is zero at x = 0 and x = L.

Solution: The first step is to write the Hamiltonian for the problem (see Section 2.4).

$$\hat{H} = -\frac{\hbar^2}{2m}\frac{d^2}{dx^2}$$

The next step is to solve for ε in terms of the adjustable parameter p (Equation 4-1).

$$\varepsilon = \frac{-\dfrac{\hbar^2 N^2}{2m}\displaystyle\int_0^L\left[\left\{\left(x - \frac{x^2}{L}\right) + p\left(x - \frac{x^2}{L}\right)^2\right\}\frac{d^2}{dx^2}\left\{\left(x - \frac{x^2}{L}\right) + p\left(x - \frac{x^2}{L}\right)^2\right\}\right]dx}{N^2\displaystyle\int_0^L\left\{\left(x - \frac{x^2}{L}\right) + p\left(x - \frac{x^2}{L}\right)^2\right\}\left\{\left(x - \frac{x^2}{L}\right) + p\left(x - \frac{x^2}{L}\right)^2\right\}dx}$$

$$\varepsilon = \frac{-\dfrac{\hbar^2}{m}\displaystyle\int_0^L\left[\left\{\left(x - \frac{x^2}{L}\right) + p\left(x - \frac{x^2}{L}\right)^2\right\}\left\{p - \frac{1}{L} - \frac{6px}{L} + \frac{6px^2}{L^2}\right\}\right]dx}{\dfrac{L^3}{30}\left[1 + \dfrac{3pL}{7} + \dfrac{p^2L^2}{21}\right]}$$

$$\varepsilon = \frac{\dfrac{\hbar^2 L}{6m}\left[1+\dfrac{2pL}{5}+\dfrac{2k^2L^2}{35}\right]}{\dfrac{L^3}{30}\left[1+\dfrac{3pL}{7}+\dfrac{p^2L^2}{21}\right]} = \frac{5\hbar^2}{mL^2}\left[\frac{1+\dfrac{2pL}{5}+\dfrac{2p^2L^2}{35}}{1+\dfrac{3pL}{7}+\dfrac{p^2L^2}{21}}\right]$$

Now the derivative of ε with respect to p is taken and set equal to zero to solve for p.

$$\frac{d\varepsilon}{dp} = 0 = 4p^2L^2 + 14pL - 21$$

The solutions are pL = 1.1331 and –4.6331. These values are now substituted into the expression for ε.

$$\varepsilon = 0.98698\left(\frac{5\hbar^2}{mL^2}\right) = \left(\frac{h}{7.9997mL^2}\right) \qquad \text{for } pL = 1.1331$$

$$\varepsilon = 10.21302\left(\frac{5\hbar^2}{mL^2}\right) = \left(\frac{h}{0.7731mL^2}\right) \qquad \text{for } pL = -4.6331$$

Since pL = 1.1331 results in a lower value for the energy of the ground-state, this value is adopted. This optimizes the trial wavefunction to:

$$\psi_{trial} = N\left[\left(x-\frac{x^2}{L}\right)+\frac{1.1331}{L}\left(x-\frac{x^2}{L}\right)^2\right].$$

The energy obtained using this trial function can now be compared to the true ground-state energy for a Particle-in-a-Box given in Equation 2-19.

True Ground-state Energy: $E_{n=1} = \dfrac{\hbar^2\pi^2}{2mL^2} = \left(\dfrac{h}{8mL^2}\right)$

Ground-state Energy from Trial Wavefunction: $E_1 = \left(\dfrac{h}{7.9997mL^2}\right)$

The optimized trial wavefunction can now be normalized (see Section 2.3).

$$\int_0^L \psi_{trial}^* \psi_{trial}\, dx = 1$$

$$N^2 \int_0^L \left[\left(x - \frac{x^2}{L} \right) + \frac{1.1331}{L} \left(x - \frac{x^2}{L} \right)^2 \right]^2 dx = 1$$

$$N = 4.4040 \sqrt{\frac{1}{L^3}}$$

$$\psi_{trial} = 4.4040 \sqrt{\frac{1}{L^3}} \left[\left(x - \frac{x^2}{L} \right) + \frac{1.1331}{L} \left(x - \frac{x^2}{L} \right)^2 \right]$$

Point of Further Understanding

The trial wavefunction in Example 4-1 (though quite good already) can be improved by adding an additional adjustable parameter.

$$\psi_{trial} = N \left\{ p_1 \left(x - \frac{x^2}{L} \right) + p_2 \left(x - \frac{x^2}{L} \right)^2 \right\}$$

List the equations that must be solved in order to optimize this wavefunction. For one additional adjustable parameter over the trial function used in Example 4-1, how many additional equations must be solved? Write the equations that must be solved to optimize this trial wavefunction.

A useful approach to obtaining a trial wavefunction is to form it from a linear combination of functions ψ_i such that the combination coefficients, c_i, become the adjustable parameters.

$$\psi_{trial} = \sum_i c_i \psi_i \tag{4-4}$$

The functions ψ_1, ψ_2, ψ_3, ... are not varied in the calculation and constitute what is called the ***basis set***. The value of ε is computed as follows:

$$\varepsilon = \frac{\langle \psi_{trial} | \hat{H} | \psi_{trial} \rangle}{\langle \psi_{trial} | \psi_{trial} \rangle} = \frac{\int \left(\sum_i c_i \psi_i \right)^* \hat{H} \left(\sum_j c_j \psi_j \right) d\tau}{\int \left(\sum_i c_i \psi_i \right)^* \left(\sum_j c_j \psi_j \right) d\tau}$$

$$\varepsilon = \frac{\sum_i \sum_j c_i^* c_j \int \psi_i^* \hat{H} \psi_j d\tau}{\sum_i \sum_j c_i^* c_j \int \psi_i^* \psi_j d\tau} = \frac{\sum_i \sum_j c_i^* c_j H_{ij}}{\sum_i \sum_j c_i^* c_j S_{ij}} \tag{4-5}$$

where $H_{ij} = \langle \psi_i | \hat{H} | \psi_j \rangle$ and $S_{ij} = \langle \psi_i | \psi_j \rangle$.

To find the minimum value of ε, Equation 4-5 is differentiated with respect to each coefficient and in turn set

$$\frac{\partial \varepsilon}{\partial c_k} = 0$$

in each case.

4.2 TIME INDEPENDENT NON-DEGENERATE PERTURBATION THEORY

The idea behind perturbation theory is that the system of interest is "perturbed" or changed slightly from a system whereby the solution is known. This can occur in two different ways: a) a new problem that has similarities to another system of which the solution is known (this happens often in chemistry) or b) the molecule or atom experiences some type of external perturbation such as a magnetic field or electromagnetic radiation (this is important in the case of spectroscopy). At this point the discussion will be limited to time-independent systems with non-degenerate quantum states. A time-independent perturbation is one in which the perturbation is not a function of time.

The Hamiltonian for the system of interest is divided into parts: the part representing system with a known solution, and then into a number of additional parts that correspond to perturbations from the known system to the system of interest.

$$\hat{H} = \hat{H}^{(0)} + \hat{H}^{(1)} + \hat{H}^{(2)} + \cdots$$

The term in the equation above with a superscript zero corresponds to the Hamiltonian for the system with a known solution (unperturbed system), $\hat{H}^{(0)}$. The rest of the terms correspond to additional terms that perturb the known system. The term $\hat{H}^{(1)}$ is a first-order perturbation, the term $\hat{H}^{(2)}$ is a second-order perturbation, and so on. The idea is that each order of perturbation is a slight change from the previous order.

Example 4-2

Problem: Consider a Particle-in-a-Box with a sinusoidal potential inside:

$$V(x) = \varepsilon \sin\left(\frac{3\pi x}{L}\right).$$

The term ε is a constant. Write the different orders of the Hamiltonian for the particle.

Solution: The complete Hamiltonian is first written for the particle.

$$\hat{H} = \hat{T} + \hat{V} = -\frac{\hbar^2}{2m}\left(\frac{\partial^2}{\partial x^2}\right) + \varepsilon \sin\left(\frac{3\pi x}{L}\right)$$

This Hamiltonian can be broken down into two parts: that of the Particle-in-a-Box Hamiltonian, $\hat{H}^{(0)}$, and that of the first-order perturbation, $\hat{H}^{(1)}$.

$$\hat{H}^{(0)} = -\frac{\hbar^2}{2m}\frac{\partial^2}{\partial x^2} \qquad\qquad \hat{H}^{(1)} = \varepsilon \sin\left(\frac{3\pi x}{L}\right)$$

The solution for the perturbed system can now be developed. The variable λ is introduced as a scalar quantity that acts as a "tunable dial" for the perturbation in the range of $0 \le \lambda \le 1$. When λ is equal to zero, there is no perturbation resulting in the unperturbed system. When the value of λ is unity, the system experiences the full perturbation. At the end of the derivation, the value of λ will be set at unity removing it from all of the expressions and the perturbation will be entirely reflected in the first and higher order perturbing Hamiltonians. The Hamiltonian for the perturbed system can be written as an expansion series in terms of λ.

$$\hat{H} = \lambda^0 \hat{H}^{(0)} + \lambda^1 \hat{H}^{(1)} + \lambda^2 \hat{H}^{(2)} + \cdots \tag{4-6}$$

The wavefunction for the system of interest at a quantum level n, ψ_n, can also be written as a sum of correction terms from the unperturbed wavefunctions, $\psi_n^{(0)}$, in an expansion series of λ.

$$\psi_n = \lambda^0 \psi_n^{(0)} + \lambda^1 \psi_n^{(1)} + \lambda^2 \psi_n^{(2)} + \cdots \tag{4-7}$$

Likewise, the energy for the perturbed system for a quantum level n can also be written as a sum of corrective terms in energy from the unperturbed system, $E_n^{(0)}$, in an expansion series of λ.

$$E_n = \lambda^0 E_n^{(0)} + \lambda^1 E_n^{(1)} + \lambda^2 E_n^{(2)} + \cdots \tag{4-8}$$

Equations 4-6 through 4-8 can now be applied to the Schroedinger equation for the problem.

$$\hat{H}\psi_n = E_n \psi_n$$

$$\left(\hat{H} - E_n\right)\psi_n = 0$$

$$[(\lambda^0 \hat{H}^{(0)} + \lambda \hat{H}^{(1)} + \lambda^2 \hat{H}^{(2)} + \cdots) - (\lambda^0 E_n^{(0)} + \lambda^1 E_n^{(1)} + \lambda^2 E_n^{(2)} + \cdots)] \\ (\lambda^0 \psi_n^{(0)} + \lambda^1 \psi_n^{(1)} + \lambda^2 \psi_n^{(2)} + \cdots) = 0 \tag{4-9}$$

In Equation 4-9, λ is the variable and the equation can be expanded and grouped in terms of orders of λ.

$$\begin{aligned} &\lambda^0 (\hat{H}^{(0)} \psi_n^{(0)} - E_n^{(0)} \psi_n^{(0)}) + \\ &\lambda^1 (\hat{H}^{(0)} \psi_n^{(1)} + \hat{H}^{(1)} \psi_n^{(0)} - E_n^{(0)} \psi_n^{(1)} - E_n^{(1)} \psi_n^{(0)}) + \\ &\lambda^2 (\hat{H}^{(0)} \psi_n^{(2)} + \hat{H}^{(1)} \psi_n^{(1)} + \hat{H}^{(2)} \psi_n^{(2)} - E_n^{(0)} \psi_n^{(1)} - E_n^{(1)} \psi_n^{(1)} - E_n^{(2)} \psi_n^{(0)}) + \\ &\cdots = 0 \end{aligned} \tag{4-10}$$

Since λ can take on any value from zero to unity, each term for the power of λ in Equation 4-10 must be individually equal to zero. Instead of just one equation, the original Schroedinger equation, there are now an infinite number of equations since the expansion in terms of powers of λ is infinite. Generally perturbation computations are only taken to the second-order and so these equations are shown below.

λ^0 terms (zero-order):

$$(\hat{H}^{(0)} - E_n^{(0)}) \psi_n^{(0)} = 0 \tag{4-11a}$$

λ^1 terms (first-order):

$$\hat{H}^{(0)} \psi_n^{(1)} + \hat{H}^{(1)} \psi_n^{(0)} - E_n^{(0)} \psi_n^{(1)} - E_n^{(1)} \psi_n^{(0)} = 0 \tag{4-11b}$$

λ^2 terms (second-order):

$$\hat{H}^{(0)} \psi_n^{(2)} + \hat{H}^{(1)} \psi_n^{(1)} + \hat{H}^{(2)} \psi_n^{(2)} - E_n^{(0)} \psi_n^{(1)} - E_n^{(1)} \psi_n^{(1)} - E_n^{(2)} \psi_n^{(0)} = 0 \tag{4-11c}$$

The reason for introducing the variable λ was to produce the separate equations 4-11a-c. Now the value of λ can be set at unity. This means that the full perturbation is reflected in the first and higher-order Hamiltonians ($\hat{H}^{(1)}, \hat{H}^{(2)}$, ...). The Equations 4-6 through 4-8 can now be rewritten with $\lambda = 1$.

$$\hat{H} = \hat{H}^{(0)} + \hat{H}^{(1)} + \hat{H}^{(2)} + \cdots \tag{4-12a}$$

$$\psi_n = \psi_n^{(0)} + \psi_n^{(1)} + \psi_n^{(2)} + \cdots \tag{4-12b}$$

$$E_n = E_n^{(0)} + E_n^{(1)} + E_n^{(2)} + \cdots \tag{4-12c}$$

The goal now is to determine the energy and wavefunction corrections for the perturbed system using Equations 4-12a-c. The most common practice is to take the corrections to second-order. Therefore, the discussion here will be limited to the first and second-order energy correction and the first-order wavefunction correction. The zero-order wavefunction and energy given in Equation 4-11a are already known as they correspond to the unperturbed system.

$$\hat{H}^{(0)}\psi_n^{(0)} = E_n^{(0)}\psi_n^{(0)}$$

To obtain the first-order energy correction for the nth quantum level of the perturbed system, $E_n^{(1)}$, Equation 4-11b is multiplied by the complex conjugate of the unperturbed wavefunction $\psi_n^{(0)*}$ and integrated over all space for the perturbed system.

$$\int \psi_n^{(0)*}[\hat{H}^{(0)}\psi_n^{(1)} + \hat{H}^{(1)}\psi_n^{(0)} - E_n^{(0)}\psi_n^{(1)} - E_n^{(1)}\psi_n^{(0)}]d\tau = 0$$

$$\int \psi_n^{(0)*}\hat{H}^{(0)}\psi_n^{(1)}d\tau + \int \psi_n^{(0)*}\hat{H}^{(1)}\psi_n^{(0)}d\tau -$$
$$E_n^{(0)}\int \psi_n^{(0)*}\psi_n^{(1)}d\tau - E_n^{(1)}\int \psi_n^{(0)*}\psi_n^{(0)}d\tau = 0$$

The above equation can be simplified by realizing that $\psi_n^{(0)}$ is orthonormal.

$$\int \psi_n^{(0)*}\hat{H}^{(0)}\psi_n^{(1)}d\tau + \int \psi_n^{(0)*}\hat{H}^{(1)}\psi_n^{(0)}d\tau - E_n^{(0)}\int \psi_n^{(0)*}\psi_n^{(1)}d\tau - E_n^{(1)} = 0$$

This equation can be further simplified by realizing that $\hat{H}^{(0)}$ is hermitian (see Section 2.5 and Equation 2-20).

$$\int \left(H^{(0)}\psi_n^{(0)}\right)^* \psi_n^{(1)}d\tau + \int \psi_n^{(0)*}\hat{H}^{(1)}\psi_n^{(0)}d\tau - E_n^{(0)}\int \psi_n^{(0)*}\psi_n^{(1)}d\tau - E_n^{(1)} = 0$$

$$E_n^{(0)}\int \psi_n^{(0)*}\psi_n^{(1)}d\tau + \int \psi_n^{(0)*}\hat{H}^{(1)}\psi_n^{(0)}d\tau - E_n^{(0)}\int \psi_n^{(0)*}\psi_n^{(1)}d\tau - E_n^{(1)} = 0$$

$$\int \psi_n^{(0)*} \hat{H}^{(1)} \psi_n^{(0)} d\tau - E_n^{(1)} = 0$$

This equation can be readily rearranged to solve for the first-order energy correction of the nth level of the perturbed system.

$$E_n^{(1)} = \int \psi_n^{(0)*} \hat{H}^{(1)} \psi_n^{(0)} d\tau = \left\langle \psi_n^{(0)} \middle| \hat{H}^{(1)} \middle| \psi_n^{(0)} \right\rangle \tag{4-13}$$

The interpretation of this result is that the first-order energy correction is a kind of average of the effect of the perturbation on the unperturbed wavefunction. The perturbation effect will be greatest at the antinodes of the wavefunction and the least at the nodes.

Example 4-3

Problem: Consider a particle in a 1-dimensional box with a potential ε in the middle 10% of the box.

Potential ε within the region of: $\frac{9L}{20} \leq x \leq \frac{11L}{20}$.

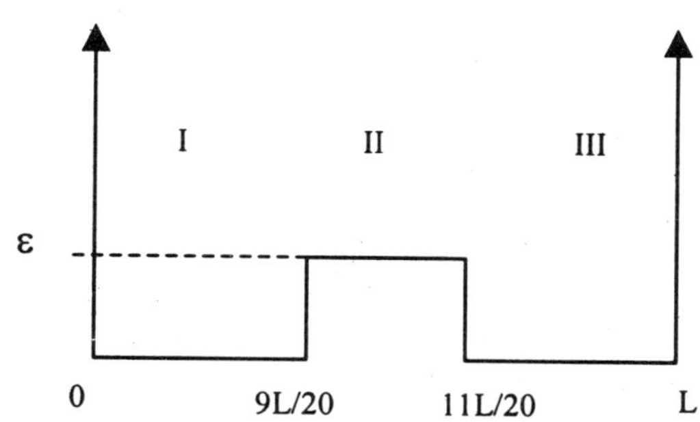

Figure 4-1. A particle in a 1-dimensional box with a constant potential blip ε in the center 10% of the box (Region II) is shown. In Regions I and III, the potential is zero as in the model 1-dimensional Particle-in-a-Box.

The potential everywhere else in the box is zero. Figure 4-1 shows a picture of this system. Calculate the energy of the system up to the first-order energy correction for the (a) ground-state, (b) first excited state, and then (c) for any level n.

Solution: As can be seen by Figure 4-1, the system is broken down into three different regions. The perturbation is isolated to Region II.

Within Region II of the box:

$$\hat{H} = -\frac{\hbar^2}{2m}\frac{d^2}{dx^2} + \varepsilon\,; \qquad \hat{H}^{(0)} = -\frac{\hbar^2}{2m}\frac{d^2}{dx^2}\,; \qquad \hat{H}^{(1)} = \varepsilon$$

$$E_n^{(0)} = \frac{n^2 h^2}{8mL^2} \qquad \text{(Equation 2-19)}$$

(a) For the ground-state (n = 1):

$$\psi_{n=1}^{(0)} = \sqrt{\frac{2}{L}}\sin\left(\frac{\pi x}{L}\right); \qquad E_{n=1}^{(0)} = \frac{h^2}{8mL^2}$$

$$E_{n=1}^{(1)} = \left\langle \psi_{n=1}^{(0)} \middle| \hat{H}^{(1)} \middle| \psi_{n=1}^{(0)} \right\rangle = \frac{2\varepsilon}{L}\int_{\frac{9L}{20}}^{\frac{11L}{20}} \sin^2\left(\frac{\pi x}{L}\right)dx = 0.1984\varepsilon$$

$$E_{n=1} = E_{n=1}^{(0)} + E_{n=1}^{(1)} = \frac{h^2}{8mL^2} + 0.1984\varepsilon$$

(b) For the first-excited state (n = 2):

$$\psi_{n=2}^{(0)} = \sqrt{\frac{2}{L}}\sin\left(\frac{2\pi x}{L}\right); \qquad E_{n=2}^{(0)} = \frac{(2)^2 h^2}{8mL^2} = \frac{h^2}{2mL^2}$$

$$E_{n=2}^{(1)} = \left\langle \psi_{n=2}^{(0)} \middle| \hat{H}^{(1)} \middle| \psi_{n=2}^{(0)} \right\rangle = \frac{2\varepsilon}{L}\int_{\frac{9L}{20}}^{\frac{11L}{20}} \sin^2\left(\frac{2\pi x}{L}\right)dx = 0.006451\varepsilon$$

$$E_{n=2} = E_{n=2}^{(0)} + E_{n=2}^{(1)} = \frac{h^2}{2mL^2} + 0.006451\varepsilon$$

(c) For any level n:

$$E_n^{(1)} = \left\langle \psi_n^{(0)} \middle| \hat{H}^{(1)} \middle| \psi_n^{(0)} \right\rangle = \frac{2\varepsilon}{L} \int_{\frac{9L}{20}}^{\frac{11L}{20}} \sin^2\left(\frac{n\pi x}{L}\right) dx$$

$$E_n^{(1)} = \varepsilon\left(\tfrac{1}{10} - (-1)^n \left(\tfrac{1}{n\pi}\right) \sin\left(\tfrac{n\pi}{10}\right)\right)$$

$$E_n = \frac{h^2 n^2}{8mL^2} + \varepsilon\left(\tfrac{1}{10} - (-1)^n \left(\tfrac{1}{n\pi}\right) \sin\left(\tfrac{n\pi}{10}\right)\right)$$

Note that the perturbation in Example 4-2 results in a greater first-order energy correction in the ground-state than in the first excited state. In fact, it can be seen by the general solution in part (c) that all odd values of n will result in a larger first-order correction than even values of n. In order to understand the reason for this behavior, the wavefunctions for the 1-dimensional Particle-in-a-Box problem, $\psi_n^{(0)}$, shown in Figure 2-2 must be compared to the perturbing potential as shown in Figure 4-1. The perturbing potential ε is limited to the center 10% of the box. All states with an even value of n have a node in the wavefunction at the center of the box in the region of the perturbing potential. As a result, the effect of the perturbing potential is minimal in even valued n states. However, in states with an odd value of n, the effect of the perturbation is the greatest because these wavefunctions have an antinode at the center of the box. The physical interpretation can be made by recalling that the square of the wavefunction for a given state n (functions are real in this case) corresponds to the probability density of the particle. States with odd values of n have minimal probability densities in the center 10% of the box whereas states with even values of n have high probability densities in this region resulting in a larger effect to the energy of the particle. The general solution in part (c) also predicts that as n increases, the first-order energy correction becomes smaller. This is because the kinetic energy of the particle increases with increasing value of n and the effect of the potential becomes less significant.

If the energy of the model or known system changes as a result of a perturbation, the wavefunction for the system also changes from its unperturbed form. The first-order correction to the wavefunction for a quantum level n for the perturbed system, $\psi_n^{(1)}$, can now be obtained. The first-order correction of the wavefunction can be expressed as a sum over the unperturbed wavefunctions.

$$\psi_n^{(1)} = \sum_k a_{kn}^{(1)} \psi_k^{(0)} \tag{4-14}$$

The sum is over all of the unperturbed wavefunctions (the basis set). The terms labeled $a_{kn}^{(1)}$ are coefficients reflecting the contribution of each of the unperturbed wavefunctions to the sum. The coefficients $a_{kn}^{(1)}$ can now be determined. Equation 4-14 is substituted into Equation 4-11b.

$$\hat{H}^{(0)}\psi_n^{(1)} + \hat{H}^{(1)}\psi_n^{(0)} - E_n^{(0)}\psi_n^{(1)} - E_n^{(1)}\psi_n^{(0)} = 0$$

$$(\hat{H}^{(0)} - E_n^{(0)})\psi_n^{(1)} + \hat{H}^{(1)}\psi_n^{(0)} - E_n^{(1)}\psi_n^{(0)} = 0$$

$$\sum_k [a_{kn}^{(1)}(\hat{H}^{(0)} - E_n^{(0)})\psi_k^{(0)}] + \hat{H}^{(1)}\psi_n^{(0)} - E_n^{(1)}\psi_n^{(0)} = 0$$

$$\sum_k [a_{kn}^{(1)}(E_k^{(0)} - E_n^{(0)})\psi_k^{(0)}] + \hat{H}^{(1)}\psi_n^{(0)} - E_n^{(1)}\psi_n^{(0)} = 0$$

This result is then multiplied by

$$\psi_i^{(0)*}$$

and integrated overall space.

$$\sum_k [a_{kn}^{(1)}(E_k^{(0)} - E_n^{(0)})\langle\psi_i^{(0)}|\psi_k^{(0)}\rangle] + \langle\psi_i^{(0)}|\hat{H}^{(1)}|\psi_n^{(0)}\rangle - E_n^{(1)}\langle\psi_i^{(0)}|\psi_n^{(0)}\rangle = 0$$

When i = n, the same result is obtained as in Equation 4-13 due to the orthonormality of $\psi^{(0)}$. When i ≠ n, all of the terms vanish except when i = k.

$$a_{kn}^{(1)}(E_k^{(0)} - E_n^{(0)}) + \left\langle \psi_k^{(0)} \middle| \hat{H}^{(1)} \middle| \psi_n^{(0)} \right\rangle = 0$$

$$a_{kn}^{(1)} = \frac{\left\langle \psi_k^{(0)} \middle| \hat{H}^{(1)} \middle| \psi_n^{(0)} \right\rangle}{E_n^{(0)} - E_k^{(0)}} \qquad (k \neq n) \qquad \text{(4-15)}$$

Equation 4-15 can now be substituted into Equation 4-14.

$$\psi_n^{(1)} = \sum_{k \neq n} \left(\frac{\left\langle \psi_k^{(0)} \middle| \hat{H}^{(1)} \middle| \psi_n^{(0)} \right\rangle}{E_n^{(0)} - E_k^{(0)}} \psi_k^{(0)} \right) \qquad \text{(4-16)}$$

The wavefunction for the perturbed system up to the first-order correction becomes the following expression.

$$\psi_n = \psi_n^{(0)} + \sum_{k \neq n} \left(\frac{\left\langle \psi_k^{(0)} \middle| \hat{H}^{(1)} \middle| \psi_n^{(0)} \right\rangle}{E_n^{(0)} - E_k^{(0)}} \psi_k^{(0)} \right) \qquad \text{(first-order correction)} \quad \text{(4-17)}$$

It is important to note that the wavefunction obtained in Equation 4-17 is not yet normalized. It will need to be normalized based on the number of unperturbed wavefunctions included in the summation. Also note that Equation 4-17 is only valid for systems with nondegenerate states (whereby $E_n^{(0)} \neq E_k^{(0)}$). The correction for degenerate states will be developed later.

Example 4-4

Problem: Determine the normalized ground-state wavefunction for the system in Example 4-3 up to the first-order correction utilizing up to n = 5 of the unperturbed wavefunctions.

Recall that:

$$\hat{H}^{(1)} = \varepsilon \text{ (within the region of } \tfrac{9L}{20} \leq x \leq \tfrac{11L}{20}\text{)}$$

and

$$\psi_n^{(0)} = \sqrt{\frac{2}{L}}\sin\left(\frac{n\pi x}{L}\right) \quad E_n^{(0)} = \frac{n^2h^2}{8mL^2}.$$

Solution: The perturbed wavefunction with a first-order correction summed to the n = 5 unperturbed wavefunction is as follows (N is the normalization constant):

$$\psi_1 \approx N(\psi_1^{(0)} + a_{21}^{(1)}\psi_2^{(0)} + a_{31}^{(1)}\psi_3^{(0)} + a_{41}^{(1)}\psi_4^{(0)} + a_{51}^{(1)}\psi_5^{(0)}).$$

Each coefficient must be solved for using Equation 4-15.

$$a_{21}^{(1)} = \frac{\frac{2\varepsilon}{L}\int_{\frac{9L}{20}}^{\frac{11L}{20}} \sin\left(\frac{2\pi x}{L}\right)\sin\left(\frac{\pi x}{L}\right)dx}{(1-2^2)E_1^{(0)}} = 0$$

$$a_{31}^{(1)} = \frac{\frac{2\varepsilon}{L}\int_{\frac{9L}{20}}^{\frac{11L}{20}} \sin\left(\frac{3\pi x}{L}\right)\sin\left(\frac{\pi x}{L}\right)dx}{(1-3^2)E_1^{(0)}} = \frac{-0.1919\varepsilon}{-8E_1^{(0)}} = 0.02399\frac{\varepsilon}{E_1^{(0)}}$$

$$a_{41}^{(1)} = \frac{\frac{2\varepsilon}{L}\int_{\frac{9L}{20}}^{\frac{11L}{20}} \sin\left(\frac{4\pi x}{L}\right)\sin\left(\frac{\pi x}{L}\right)dx}{(1-4^2)E_1^{(0)}} = 0$$

$$a_{51}^{(1)} = \frac{\frac{2\varepsilon}{L}\int_{\frac{9L}{20}}^{\frac{11L}{20}} \sin\left(\frac{5\pi x}{L}\right)\sin\left(\frac{\pi x}{L}\right)dx}{(1-5^2)E_1^{(0)}} = \frac{0.1794\varepsilon}{-24E_1^{(0)}} = -0.007475\frac{\varepsilon}{E_1^{(0)}}$$

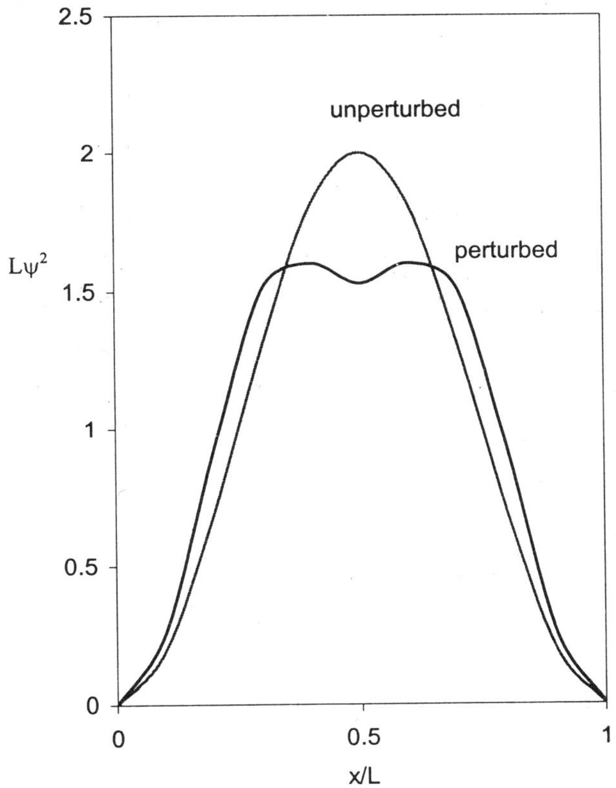

Figure 4-2. The ground-state perturbed wavefunction up to a first-order correction for a particle in a 1-dimensional box with a positive potential blip in the center 10% of the box as described in Examples 4-3 and 4-4. The potential ε is equal to four times the unpertubed ground-state energy.

Each coefficient is dimensionless as required since ε represents a potential. Note that each even value of n results in a coefficient equal to zero representing no contribution from these wavefunctions. An exact representation of the perturbed wavefunction is obtained by adding an infinite number wavefunctions; however, the contribution of each function decreases with higher-order wavefunctions. The wavefunction for the perturbed system becomes:

$$\psi_1 \approx N\left(\psi_1^{(0)} + 0.02399\frac{\varepsilon}{E_1^{(0)}}\psi_3^{(0)} - 0.007475\frac{\varepsilon}{E_1^{(0)}}\psi_5^{(0)}\right).$$

The wavefunction can now be normalized.

$$N^2\int_0^L\left(\psi_1^{(0)} + 0.02399\frac{\varepsilon}{E_1^{(0)}}\psi_3^{(0)} - 0.007475\frac{\varepsilon}{E_1^{(0)}}\psi_5^{(0)}\right)^2 dx = 1$$

$$N = \sqrt{\frac{1}{1+0.0006314\frac{\varepsilon}{E_1^{(0)}}}} \cong 1$$

The normalized wavefunction becomes as follows:

$$\psi_1 \approx \psi_1^{(0)} + 0.02399\frac{\varepsilon}{E_1^{(0)}}\psi_3^{(0)} - 0.007475\frac{\varepsilon}{E_1^{(0)}}\psi_5^{(0)}.$$

For a positive potential ε, the effect of the first-order correction to the ground-state wavefunction, $\psi_1^{(1)}$, is to reduce the probability density of the particle in the center of the box. This is shown in Figure 4-2 for the case where $\varepsilon = 4E_1^{(0)}$. The probability density has diminished for the particle in the center 10% of the box where the positive potential blip ε exists.

Point of Further Understanding

Use a spreadsheet program to explore the first-order ground-state wavefunction obtained in Example 4-4. In the first column, list x/L values from 0 to 1 in 0.1 increments. In the second column, compute the value of the ground-state unperturbed wavefunction ($\psi_1^{(0)}\sqrt{L}$) for each of the x/L values in the first column. Then create a defined cell representing it as the ratio of

$$\frac{\varepsilon}{E_1^{(0)}}.$$

Initially set this ratio to 4. In the third column of the spreadsheet, compute the value of the ground-state perturbed wavefunction as given in Example 4-4 for each of the x/L values in the first column using the defined cell. Make a plot that contains both the perturbed and unperturbed wavefunction as a function of x/L. Change the value of the defined cell

$$\frac{\varepsilon}{E_1^{(0)}}$$

to values less than 1 to very large values observing how the perturbed wavefunction changes in shape. Change the value of the defined cell to a negative ratio signifying a negative potential. How does this change the shape of the perturbed wavefunction? Is this shape of the perturbed wavefunction expected? Explain.

The second-order correction to the nth state energy of the perturbed system, $E_n^{(2)}$, can now be developed. The approach is similar to obtaining the first-order energy correction. Equation 4-11c is multiplied by the complex conjugate of the unperturbed nth state wavefunction, $\psi_n^{(0)}$, and integrated over the perturbed system.

$$\int \psi_n^{(0)*} (\hat{H}^{(0)}\psi_n^{(2)} + \hat{H}^{(1)}\psi_n^{(1)} + \hat{H}^{(2)}\psi_n^{(2)} \\ - E_n^{(0)}\psi_n^{(1)} - E_n^{(1)}\psi_n^{(1)} - E_n^{(2)}\psi_n^{(0)})d\tau = 0$$

Upon taking advantage of the orthonormality of the $\psi^{(0)}$ wavefunctions and the hermiticity of $\hat{H}^{(0)}$, the following expression results for the second-order correction for the n state of the perturbed system.

$$E_n^{(2)} = \int \psi_n^{(0)*} \hat{H}^{(2)} \psi_n^{(0)} d\tau + \sum_{k \neq n} \frac{(\int \psi_n^{(0)*} \hat{H}^{(1)} \psi_k^{(0)} d\tau)(\int \psi_k^{(0)*} \hat{H}^{(1)} \psi_n^{(0)} d\tau)}{E_n^{(0)} - E_k^{(0)}} \quad (4\text{-}18)$$

In terms of Dirac notation, Equation 4-18 may be expressed as follows:

$$E_n^{(2)} = \left\langle \psi_n^{(0)} \middle| \hat{H}^{(2)} \middle| \psi_n^{(0)} \right\rangle + \sum_{k \neq n} \frac{\left\langle \psi_n^{(0)} \middle| \hat{H}^{(1)} \middle| \psi_k^{(0)} \right\rangle \left\langle \psi_k^{(0)} \middle| \hat{H}^{(1)} \middle| \psi_n^{(0)} \right\rangle}{E_n^{(0)} - E_k^{(0)}}. \quad (4\text{-}19)$$

The expressions in Equations 4-18 and 4-19 are valid only for non-degenerate systems ($E_n^{(0)} \neq E_k^{(0)}$). Also note that it is possible to have a non-zero second-order energy correction (and even higher orders) even if the perturbed system has only a first-order perturbing Hamiltonian. While the first-order correction to the energy represents an "average" of the perturbation to a given unperturbed state, the second-order correction to the energy represents the "mixing" between unperturbed states as a result of the perturbation.

Example 4-5

Problem: Consider the same system as in Example 4-2, a Particle-in-a-Box with a sinusoidal potential inside,

$$\hat{H}^{(1)} = \varepsilon \sin\left(\frac{3\pi x}{L}\right).$$

Calculate the second-order energy correction to the ground-state of the perturbed system including up to the n = 7 unperturbed wavefunction.

Solution: Since the perturbed system only has a first-order perturbation, $\hat{H}^{(2)} = 0$. Equation 4-18 summed to n = 7 is as follows:

$$E_1^{(2)} = \frac{(\int \psi_1^{(0)*} \hat{H}^{(1)} \psi_2^{(0)} d\tau)(\int \psi_2^{(0)*} \hat{H}^{(1)} \psi_1^{(0)} d\tau)}{E_1^{(0)} - E_2^{(0)}}$$

$$+ \frac{(\int \psi_1^{(0)*} \hat{H}^{(1)} \psi_3^{(0)} d\tau)(\int \psi_3^{(0)*} \hat{H}^{(1)} \psi_1^{(0)} d\tau)}{E_1^{(0)} - E_3^{(0)}}$$

$$+ \frac{(\int \psi_1^{(0)*} \hat{H}^{(1)} \psi_4^{(0)} d\tau)(\int \psi_4^{(0)*} \hat{H}^{(1)} \psi_1^{(0)} d\tau)}{E_1^{(0)} - E_4^{(0)}}$$

$$+ \frac{(\int \psi_1^{(0)*} \hat{H}^{(1)} \psi_5^{(0)} d\tau)(\int \psi_5^{(0)*} \hat{H}^{(1)} \psi_1^{(0)} d\tau)}{E_1^{(0)} - E_5^{(0)}}$$

$$\frac{(\int \psi_1^{(0)*} \hat{H}^{(1)} \psi_6^{(0)} d\tau)(\int \psi_6^{(0)*} \hat{H}^{(1)} \psi_1^{(0)} d\tau)}{E_1^{(0)} - E_6^{(0)}}$$

$$+ \frac{(\int \psi_1^{(0)*} \hat{H}^{(1)} \psi_7^{(0)} d\tau)(\int \psi_7^{(0)*} \hat{H}^{(1)} \psi_1^{(0)} d\tau)}{E_1^{(0)} - E_7^{(0)}}.$$

Since the $\psi^{(0)}$ wavefunctions are real and $\hat{H}^{(1)}$ is a sine function,

$$\int \psi_n^{(0)} \hat{H}^{(1)} \psi_k^{(0)} d\tau = \int \psi_k^{(0)} \hat{H}^{(1)} \psi_n^{(0)} d\tau .$$

$$E_1^{(2)} = \frac{(\int \psi_1^{(0)} \hat{H}^{(1)} \psi_2^{(0)} d\tau)^2}{E_1^{(0)} - E_2^{(0)}} + \frac{(\int \psi_1^{(0)} \hat{H}^{(1)} \psi_3^{(0)} d\tau)^2}{E_1^{(0)} - E_3^{(0)}}$$

$$+ \frac{(\int \psi_1^{(0)} \hat{H}^{(1)} \psi_4^{(0)} d\tau)^2}{E_1^{(0)} - E_4^{(0)}} + \frac{(\int \psi_1^{(0)} \hat{H}^{(1)} \psi_5^{(0)} d\tau)^2}{E_1^{(0)} - E_5^{(0)}}$$

$$+ \frac{(\int \psi_1^{(0)} \hat{H}^{(1)} \psi_6^{(0)} d\tau)^2}{E_1^{(0)} - E_6^{(0)}} + \frac{(\int \psi_1^{(0)} \hat{H}^{(1)} \psi_7^{(0)} d\tau)^2}{E_1^{(0)} - E_7^{(0)}}$$

The integrals in the numerators can now be solved. Due to symmetry, only odd values of n will contribute to the sum.

$$\int \psi_1^{(0)} \hat{H}^{(1)} \psi_n^{(0)} d\tau$$

$$= \frac{2\varepsilon}{L} \int_0^L \sin\left(\frac{\pi x}{L}\right) \sin\left(\frac{3\pi x}{L}\right) \sin\left(\frac{n\pi x}{L}\right) dx = 3.8197\varepsilon \frac{n[\cos(n\pi) - 1]}{(n^4 - 20n^2 + 64)}$$

n	3	5	7
$\dfrac{\int \psi_1^{(0)} \hat{H}^{(1)} \psi_n^{(0)} d\tau}{\varepsilon}$	0.6548	-0.2021	-0.0361

Since $E_1^{(0)} - E_n^{(0)} = (1-n^2)E_1^{(0)}$, the second-order energy correction for the ground-state of the perturbed system becomes:

$$E_1^{(2)} \approx \left(\frac{0.6548^2}{1-3^2} + \frac{(-0.2021)^2}{1-5^2} + \frac{(-0.0361)^2}{1-7^2} \right) \frac{\varepsilon^2}{E_1^{(0)}} = -0.05532 \frac{\varepsilon^2}{E_1^{(0)}}$$

4.3 TIME-INDEPENDENT DEGENERATE PERTURBATION THEORY

The non-degenerate perturbation expressions that have been developed in the previous section will result in zero denominators for degenerate systems. As a result, the approach used to obtain the expressions for the various orders of corrections for a degenerate system must be modified from the approach used in non-degenerate systems.

Consider an unperturbed system where a given energy level n has an r-fold degeneracy. This means that there are r wavefunctions that will result in the same energy, $E_n^{(0)}$, when applied to the Hamiltonian, $\hat{H}^{(0)}$. In this notation scheme, the n refers to the various degenerate states ($1 \le n \le r$).

$$\hat{H}^{(0)} \psi_n^{(0)} = E_n^{(0)} \psi_n^{(0)} \tag{4-20}$$

$$E_1^{(0)} = E_2^{(0)} = \cdots = E_r^{(0)} \tag{4-21}$$

Now suppose this degenerate system experiences a perturbation. The Hamiltonian for this perturbed system is $\hat{H}$, and the wavefunctions for the perturbed system are ψ_n. The wavefunctions ψ_n may be non-degenerate, have a fraction of the degeneracy, or in some cases no change in the degeneracy relative to the unperturbed system. The change in degeneracy in the perturbed system has to do with the nature and symmetry of the perturbation relative to the unperturbed wavefunctions. For the perturbed system:

$$\hat{H} = \lambda^0 \hat{H}^{(0)} + \lambda^1 \hat{H}^{(1)} + \cdots \qquad (4\text{-}22)$$

$$\hat{H}\psi_n = E_n \psi_n . \qquad (4\text{-}23)$$

Equation 4-22 for the perturbed degenerate system is analogous to Equation 4-6 for the non-degenerate perturbed system. A natural assumption at this point is that the zero-order wavefunction of ψ_n is $\psi_n^{(0)}$ as in Equation 4-20. If the eigenvalue $E_n^{(0)}$ is non-degenerate, then the assumption is certainly true as there is a unique normalized eigenfunction that will satisfy Equation 4-20 (this is precisely the approach used in the previous section). However, if $E_n^{(0)}$ has a degeneracy of r, then there are r normalized eigenfunctions along with an infinite number of other normalized linear combinations of the r functions that will satisfy Equation 4-20.

$$c_1\psi_1^{(0)} + c_2\psi_2^{(0)} + \cdots + c_r\psi_r^{(0)}$$
$$c_1'\psi_1^{(0)} + c_2'\psi_2^{(0)} + \cdots + c_r'\psi_r^{(0)}$$
$$\vdots$$

For the unperturbed system, any normalized linear combinations of the r unperturbed wavefunctions are acceptable solutions; however for the perturbed system, only certain normalized linear combinations form the correct zero-order perturbed (unperturbed) wavefunctions $\phi_n^{(0)}$.

$$\phi_n^{(0)} = \sum_{i=1}^{r} c_i \psi_i^{(0)} \qquad (1 \le i \le r) \qquad (4\text{-}24)$$

The $\phi_n^{(0)}$ wavefunctions depend on the type of perturbation that the system experiences. The perturbed energy eigenvalues and wavefunctions can now be written in terms of orders of λ:

$$\psi_n = \lambda^0 \phi_n^{(0)} + \lambda^1 \psi_n^{(1)} + \cdots \qquad (n = 1, 2, 3, ..., r) \qquad (4\text{-}25)$$

$$E_n = \lambda^0 E_n^{(0)} + \lambda^1 E_n^{(1)} + \cdots \qquad (n = 1, 2, 3, ..., r). \qquad (4\text{-}26)$$

For an r-degenerate system, there will be r perturbed wavefunctions, ψ_n, and r energy eigenvalues, E_n. Equations 4-22, 4-25, and 4-26 can now be substituted into the Schroedinger equation for the perturbed system (Equation 4-23).

$$\hat{H}\psi_n = E_n \psi_n$$

$$(\lambda^0 \hat{H}^{(0)} + \lambda^1 \hat{H}^{(1)} + \cdots)(\lambda^0 \phi_n^{(0)} + \lambda^1 \psi_n^{(0)} + \cdots)$$
$$= (\lambda^0 E_n^{(0)} + \lambda^1 E_n^{(1)} + \cdots)(\lambda^0 \phi_n^{(0)} + \lambda^1 \psi_n^{(0)} + \cdots)$$

Ordering the orders of the coefficients of λ yields the following:

λ^0 (zero-order): $$\hat{H}^{(0)}\phi_n^{(0)} = E_n^{(0)}\phi_n^{(0)} \qquad (n = 1, 2, ..., r) \qquad (4\text{-}27)$$

λ^1 (first-order): $$\hat{H}^{(0)}\psi_n^{(1)} + \hat{H}^{(1)}\phi_n^{(0)} = E_n^{(0)}\psi_n^{(1)} + E_n^{(1)}\varphi_n^{(0)}$$
$$(\hat{H}^{(0)} - E_n^{(0)})\psi_n^{(1)} = (\hat{H}^{(1)} - E_n^{(1)})\phi_n^{(0)} \qquad (n = 1, 2, ..., r). \qquad (4\text{-}28)$$

The same procedure as in the non-degenerate case is now continued to obtain the first-order correction to the energy. Equation 4-28 is now multiplied by the complex conjugate of one the unperturbed degenerate wavefunctions, $\psi_m^{(0)*}$ ($1 \le m \le r$), and integrated over all space.

$$\left\langle \psi_m^{(0)} \middle| \hat{H}^{(0)} \middle| \psi_n^{(1)} \right\rangle - E_n^{(0)} \left\langle \psi_m^{(0)} \middle| \psi_n^{(1)} \right\rangle = \left\langle \psi_m^{(0)} \middle| \hat{H}^{(1)} \middle| \phi_n^{(0)} \right\rangle - E_n^{(1)} \left\langle \psi_m^{(0)} \middle| \phi_n^{(0)} \right\rangle$$

Since $\hat{H}^{(0)}$ is hermitian (see Section 4.2),

$$\left\langle \psi_m^{(0)} \middle| \hat{H}^{(0)} \middle| \psi_n^{(1)} \right\rangle = E_n^{(0)} \left\langle \psi_m^{(0)} \middle| \psi_n^{(1)} \right\rangle.$$

$$E_n^{(0)} \left\langle \psi_m^{(0)} \middle| \psi_n^{(1)} \right\rangle - E_n^{(0)} \left\langle \psi_m^{(0)} \middle| \psi_n^{(1)} \right\rangle = \left\langle \psi_m^{(0)} \middle| \hat{H}^{(1)} \middle| \phi_n^{(0)} \right\rangle - E_n^{(1)} \left\langle \psi_m^{(0)} \middle| \phi_n^{(0)} \right\rangle$$

$$\left\langle \psi_m^{(0)} \middle| \hat{H}^{(1)} \middle| \phi_n^{(0)} \right\rangle - E_n^{(1)} \left\langle \psi_m^{(0)} \middle| \phi_n^{(0)} \right\rangle = 0 \qquad (4\text{-}29)$$

The expression for $\phi_n^{(0)}$ in Equation 4-24 can now be substituted into Equation 4-29.

$$\sum_{i=1}^{r} c_i \left\langle \psi_m^{(0)} \middle| \hat{H}^{(1)} \middle| \psi_i^{(0)} \right\rangle - E_n^{(1)} \sum_{i=1}^{r} c_i \left\langle \psi_m^{(0)} \middle| \psi_i^{(0)} \right\rangle = 0 \qquad (4\text{-}30)$$

The degenerate unperturbed wavefunctions $\psi_i^{(0)}$ can always be chosen to be orthonormal. As a result, the relationship for orthonormality can be employed (see Equation 2-24).

$$\left\langle \psi_m^{(0)} \middle| \psi_i^{(0)} \right\rangle = \delta_{mi} \qquad \begin{aligned} \delta_{mi} &= 1 \text{ (if i = m)} \\ &= 0 \text{ (if i} \neq \text{m)} \end{aligned}$$

The orthonormality relationship can now substituted into Equation 4-30.

$$\begin{aligned} &\sum_{i=1}^{r} c_i \left\langle \psi_m^{(0)} \middle| \hat{H}^{(1)} \middle| \psi_i^{(0)} \right\rangle - E_n^{(1)} \sum_{i=1}^{r} c_i \delta_{mi} = 0 \\ &\sum_{i=1}^{r} c_i \left(\left\langle \psi_m^{(0)} \middle| \hat{H}^{(1)} \middle| \psi_i^{(0)} \right\rangle - E_n^{(1)} \delta_{mi} \right) = 0 \end{aligned} \qquad (1 \le i \le r) \qquad (4\text{-}31)$$

The expression in Equation 4-31 results in r homogeneous equations.

$$c_1 \left(\left\langle \psi_1^{(0)} \middle| \hat{H}^{(1)} \middle| \psi_1^{(0)} \right\rangle - E_n^{(1)} \right) + c_2 \left\langle \psi_1^{(0)} \middle| \hat{H}^{(1)} \middle| \psi_2^{(0)} \right\rangle + \cdots + c_r \left\langle \psi_1^{(0)} \middle| \hat{H}^{(1)} \middle| \psi_r^{(0)} \right\rangle = 0$$

$$c_1 \left\langle \psi_2^{(0)} \middle| \hat{H}^{(1)} \middle| \psi_1^{(0)} \right\rangle + c_2 \left(\left\langle \psi_2^{(0)} \middle| \hat{H}^{(1)} \middle| \psi_2^{(0)} \right\rangle - E_n^{(1)} \right) + \cdots + c_r \left\langle \psi_2^{(0)} \middle| \hat{H}^{(1)} \middle| \psi_r^{(0)} \right\rangle = 0$$

$$\vdots$$

$$c_1 \left\langle \psi_r^{(0)} \middle| \hat{H}^{(1)} \middle| \psi_1^{(0)} \right\rangle + c_2 \left\langle \psi_r^{(0)} \middle| \hat{H}^{(1)} \middle| \psi_2^{(0)} \right\rangle + \cdots + c_r \left(\left\langle \psi_r^{(0)} \middle| \hat{H}^{(1)} \middle| \psi_r^{(0)} \right\rangle - E_n^{(1)} \right) = 0$$

In order for these equations to have a non-trivial solution, the determinant of the coefficients must be equal to zero. The resulting expression is called the ***secular equation***.

$$\det\left|\left\langle\psi_m^{(0)}\right|\hat{H}^{(1)}\left|\psi_i^{(0)}\right\rangle - E_n^{(1)}\delta_{mi}\right| = 0 \tag{4-32}$$

The first-order correction to the energy is determined by solving the determinant. The coefficients c_i can then be found by substituting back into Equation 4-31.

Example 4-6

Problem: The Particle-on-a-Ring experiences the following potential:

$$V(x) = \varepsilon \sin^2\phi \ .$$

Determine the first-order energy corrections to the degenerate $m_l = \pm 1$ states and the value of the coefficients for the zero-order wavefunctions, $\phi_n^{(0)}$.

Solution: The first-order perturbing Hamiltonian for the problem can be written for convenience in the following form:

$$\hat{H}^{(1)} = \varepsilon \sin^2\phi = \varepsilon\left(\frac{2 - e^{2i\phi} - e^{-2i\phi}}{4}\right).$$

The unperturbed wavefunctions for the Particle-on-a-Ring are given as (see Equation 3-7):

$$\psi_{m_l}^{(0)} = \sqrt{\frac{1}{2\pi}}e^{im_l\phi} \ .$$

The following equations must be solved:

$$\begin{aligned} c_1\left(\left\langle\psi_1^{(0)}\right|\hat{H}^{(1)}\left|\psi_1^{(0)}\right\rangle - E_n^{(1)}\right) + c_2\left\langle\psi_1^{(0)}\right|\hat{H}^{(1)}\left|\psi_{-1}^{(0)}\right\rangle = 0 \\ c_1\left\langle\psi_{-1}^{(0)}\right|\hat{H}^{(1)}\left|\psi_1^{(0)}\right\rangle + c_2\left(\left\langle\psi_{-1}^{(0)}\right|\hat{H}^{(1)}\left|\psi_{-1}^{(0)}\right\rangle - E_n^{(1)}\right) = 0 \end{aligned} \ .$$

The following integral needs to be solved:

$$\left\langle \psi_{m_l'}^{(0)} \middle| \hat{H}^{(1)} \middle| \psi_{m_l}^{(0)} \right\rangle = \frac{\varepsilon}{2\pi} \int_0^{2\pi} e^{-im_l'\phi} \left(\frac{2 - e^{2i\phi} - e^{-2i\phi}}{4} \right) e^{im_l\phi}\, d\phi$$

$$= \frac{\varepsilon}{8\pi} \left(\int_0^{2\pi} 2e^{i(m_l - m_l')\phi} d\phi - \int_0^{2\pi} e^{i(2+m_l-m_l')} d\phi - \int_0^{2\pi} e^{i(m_l - m_l' - 2)} d\phi \right)$$

$$= \frac{\varepsilon}{8\pi} \left(4\pi\delta_{m_l' m_l} - 2\pi\delta_{m_l', m_l+2} - 2\pi\delta_{m_l', m_l-2} \right).$$

This results in the following equations:

$$c_1\left(\frac{\varepsilon}{2} - E_n^{(1)}\right) + c_2\left(\frac{-\varepsilon}{4}\right) = 0$$

$$c_1\left(\frac{-\varepsilon}{4}\right) + c_2\left(\frac{\varepsilon}{2} - E_n^{(1)}\right) = 0$$

The non-trivial solution is when the following determinant is equal to zero.

$$\begin{vmatrix} \left(\frac{\varepsilon}{2} - E_n^{(1)}\right) & \frac{-\varepsilon}{4} \\ \frac{-\varepsilon}{4} & \left(\frac{\varepsilon}{2} - E_n^{(1)}\right) \end{vmatrix} = 0$$

$$\left(\frac{\varepsilon}{2} - E_n^{(1)}\right)^2 - \left(\frac{\varepsilon}{4}\right)^2 = 0; \qquad E_n^{(1)} = \frac{\varepsilon}{2} \pm \frac{\varepsilon}{4}$$

This results in the following first-order energy corrections:

$$E_+^{(1)} = \frac{3\varepsilon}{4} \qquad \text{and} \qquad E_-^{(1)} = \frac{\varepsilon}{4}.$$

Now all that remains is to determine the value of the coefficients for $\phi_n^{(0)}$ for the two resulting states. Since $\phi_n^{(0)}$ must be normalized, one equation that must be satisfied is

$$|c_1|^2 + |c_2|^2 = 1.$$

For the $E_+^{(1)} = \frac{3\varepsilon}{4}$ case:

$$\begin{matrix} c_1\left(\frac{-\varepsilon}{4}\right) + c_2\left(\frac{-\varepsilon}{4}\right) = 0 \\ c_1\left(\frac{-\varepsilon}{4}\right) + c_2\left(\frac{-\varepsilon}{4}\right) = 0 \end{matrix} ; \quad c_1 = -c_2 = \sqrt{\frac{1}{2}}; \quad \phi_+^{(0)} = \frac{\psi_1^{(0)} - \psi_{-1}^{(0)}}{\sqrt{2}}.$$

For the $E_-^{(1)} = \frac{\varepsilon}{4}$ case:

$$\begin{matrix} c_1\left(\frac{\varepsilon}{4}\right) + c_2\left(\frac{-\varepsilon}{4}\right) = 0 \\ c_1\left(\frac{-\varepsilon}{4}\right) + c_2\left(\frac{\varepsilon}{4}\right) = 0 \end{matrix} ; \quad c_1 = c_2 = \sqrt{\frac{1}{2}}; \quad \phi_-^{(0)} = \frac{\psi_1^{(0)} + \psi_{-1}^{(0)}}{\sqrt{2}}.$$

As a result of the perturbation, the degeneracy is lost for the $m_l = \pm 1$ state for the Particle-on-a-Ring.

PROBLEMS AND EXERCISES

4.1) Determine the normalized and optimized ground-state wavefunction for the one-dimensional Particle-in-a-Box using the following trial wavefunction:

$$\psi_{trial} = N\left(\left(x - \frac{x^2}{L}\right) + p\left(x - \frac{x^2}{L}\right)^3\right)$$

where p is an adjustable parameter and N is the normalization constant. Compare the ground-state energy obtained using this trial function to the true ground-state energy.

4.2) A simple model for predicting ultra-violet/visible absorption spectra of conjugated polyenes is the free-electron molecular orbital model. In this model, the electrons of the conjugated π system are free to travel the length of the conjugated carbon chain. In this model, the 1-dimensional particle-in-a-box model problem can be used to simulate the molecular orbitals of the electrons in the conjugated chain. The length of the box, L, is equal to the length of the conjugated carbon chain.

$$L = n_{CC}(150\,pm)$$

The value n_{cc} corresponds to the number of carbon/carbon bonds in the conjugated chain. The electrons can be ordered two at a time in each energy level. In this problem, we will consider lycopene.

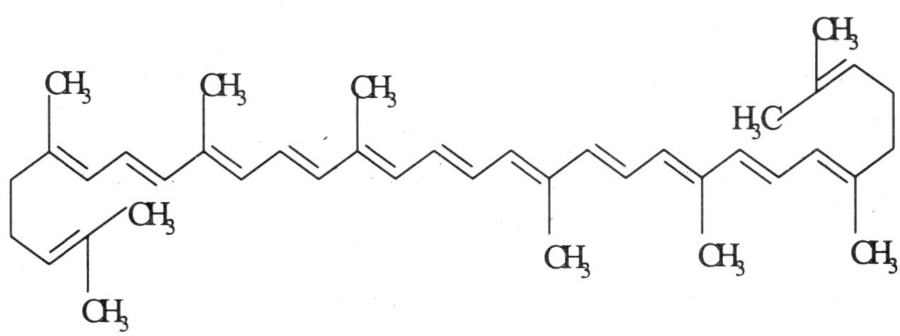

a. Determine the energy of the transition from the lowest occupied to the first unoccupied energy level for lycopene. Determine the wavelength needed for this transition ($\lambda = hc/\Delta E$). The experimental value for lycopene is 474 nm.

b. The value obtained from part (a) above is not in good agreement with what is obtained experimentally. This is in part because the potential on the electron is not zero or constant. The potential can be improved by having it change sinusoidally along the polyene chain. Choose a sine function that will have an appropriate periodicity over lycopene and treat it as a first-order perturbation. Repeat the computations from part (a) using up to a second-order energy correction.

4.3) Consider a Particle-in-a-Box with a ramp-like potential that increases from x = 0 to x = L.

$$V(x) = \varepsilon x$$

a. Determine up to the second-order energy correction along with a first-order wavefunction correction for the n = 1 and n = 2 states.

b. Suppose the potential in part (a) is due to gravity because the box is vertical. Apply the results from part (a) for an electron at the surface of the Earth.

4.4) Determine the first-order energy corrections for the $m_l = \pm 2$ states for the Particle-on-a-Ring problem when

$$\hat{H}^{(1)} = \varepsilon \cos^2 \phi \, .$$

Find the values of the zero-order wavefunction coefficients.

Chapter 5

Particles Encountering a Finite Potential Energy

The systems we have studied so far have all involved particles confined to a *limited* region of space such as a box, circle, or a sphere via infinite potentials. In the case of a box, it is explicitly stated that the potentials at the "walls" of the box are infinite. Though not explicitly stated, in order to confine a particle to a circle or a sphere of constant radius r, the potential must be infinite for anyplace outside the circle or sphere. Removing the requirement of infinite potential at a given point has the effect of no longer being able to completely contain a particle into any limited region of space even if the potential exceeds the energy of the particle. In addition, quantum mechanics predicts that a particle can be reflected by a potential even though its energy is in excess of that potential. These phenomena are entirely a quantum mechanical result due to the wave nature of matter.

5.1 HARMONIC OSCILLATOR

The harmonic oscillator is used as a simple model for the vibrational motion of atoms along bonds in molecules. This will in turn be used to model infrared absorption spectroscopy in the next chapter.

Consider a spring with a force constant k with a Hooke's law potential anchored at one end and attached to a mass of m on the other end as shown in Figure 1-1. The particle is confined to travel only along the x-coordinate, and for convenience, the equilibrium position of the spring (point of zero potential energy) is at $x_0 = 0$. The Hamiltonian for the particle becomes:

$$\hat{H} = \hat{T} + \hat{V} = \frac{\hat{p}^2}{2m} + \tfrac{1}{2}kx^2 = -\frac{\hbar^2}{2m}\left(\frac{d^2}{dx^2}\right) + \tfrac{1}{2}kx^2 \tag{5-1}$$

The Schroedinger equation for this system can now be readily written.

$$\hat{H}\psi(x) = E\psi(x)$$

$$-\frac{\hbar^2}{2m}\left(\frac{d^2\psi(x)}{dx^2}\right) + \tfrac{1}{2}kx^2\psi(x) = E\psi(x) \tag{5-2}$$

The solution to this differential equation is well known. The energy eigenvalues are quantized as follows:

$$E_\nu = (\nu + \tfrac{1}{2})\hbar\omega\,; \qquad \nu = 0, 1, 2, \ldots \tag{5-3}$$

and

$$\omega = \sqrt{\frac{k}{m}} \;\; (radians/sec). \tag{5-4}$$

The wavefunctions can be expressed in the following manner:

$$\psi_\nu(x) = \psi_\nu\left(\tfrac{z}{c}\right) = N_\nu h_\nu(z)e^{-z^2/2} \tag{5-5}$$

$$z = cx \tag{5-6}$$

$$c = \sqrt{\frac{m\omega}{\hbar}} \tag{5-7}$$

$$N_\nu = \sqrt{\left(\frac{c}{2^\nu \nu! \pi^{\frac{1}{2}}}\right)} \tag{5-8}$$

The functions $h_\nu(z)$ in Equation 5-5 are polynomials in z known as the Hermite polynomials. The Hermite polynomials can be generated from the following formula:

$$h_v(z) = (-1)^v e^{z^2} \frac{d^v}{dz^v} e^{-z^2} . \tag{5-9}$$

The Hermite polynomials for the $v+1$ state (the next state) can also be obtained from the following recursion relationship given that $h_0(z) = 1$.

$$h_{v+1}(z) = 2zh_v(z) - \frac{dh_v(z)}{dz} \tag{5-10}$$

The first six Hermite polynomials are shown in Table 5-1.

The normalization constant for the wavefunction is obtained by integrating the wavefunction squared (the wavefunction is real for the harmonic oscillator) over all space. As mentioned previously, only an infinite potential can completely contain a particle in a limited region of space. Since the potential for the system approaches infinity at the limit that x approaches infinity, the limits on the integration overall space must be $-\infty \le x \le \infty$.

Table 5-1. The Hermite polynomials are tabulated up to v = 5.

v	*Hermite Polynomials, $h_v(z)$*
0	1
1	$2z$
2	$4z^2 - 2$
3	$8z^3 - 12z$
4	$16z^4 - 48z^2 + 12$
5	$32z^5 - 160z^3 + 120z$

Example 5-1

Problem: Demonstrate that the ground-state harmonic oscillator wavefunction is (a) normalized and (b) is orthogonal with the first excited state.

Solution:

(a) The ground-state wavefunction, ψ_0, is written as follows:

$$\psi_0 = N_0 h_0(z) e^{-z^2/2} = \sqrt{\frac{c}{\pi^{\frac{1}{2}}}} e^{-z^2/2}$$

If the wavefunction is normalized, the following integral must be equal to one.

$$\int_{-\infty}^{\infty} \psi_0^* \psi_0 dx = \frac{c}{\sqrt{\pi}} \int_{-\infty}^{\infty} e^{-z^2} dx = \frac{2c}{\sqrt{\pi}} \int_0^{\infty} e^{-c^2x^2} dx = \frac{2c}{\sqrt{\pi}} \left(\frac{\sqrt{\pi}}{2c} \right) = 1$$

This confirms that the ground-state wavefunction is normalized. Based on this result, it is extrapolated that the rest of the wavefunctions are all normalized though the normalization constant is different for each level of ν.

(b) The wavefunction, ψ_1, can be written as follows:

$$\psi_1 = \sqrt{\frac{c}{2\pi^{\frac{1}{2}}}} (2z) e^{-z^2/2}$$

If the wavefunctions are orthogonal, then the following integral is equal to zero.

$$\int_{-\infty}^{\infty} \psi_1^* \psi_0 dx = c\sqrt{\frac{2}{\pi}} \int_{-\infty}^{\infty} z e^{-z^2} dx = c\sqrt{\frac{2}{\pi}} \left(\int_{-\infty}^{0} cx e^{-c^2x^2} dx + \int_0^{\infty} cx e^{-c^2x^2} dx \right) = 0$$

This demonstrates that the wavefunctions ψ_0 and ψ_1 are orthogonal. This is true for all of the harmonic oscillator wavefunctions.

Example 5-2

Problem: Determine explicitly the energy eigenvalue for the ground-state of the harmonic oscillator by using the Schroedinger equation.

Solution:

$$-\frac{\hbar^2}{2m}\left(\frac{d^2\psi_0}{dx^2}\right)+\tfrac{1}{2}kx^2\psi_0=E_0\psi_0$$

$$-\frac{\hbar^2}{2m}\left(-c^2N_0e^{-c^2x^2/2}+c^4x^2N_0e^{-c^2x^2/2}\right)+\tfrac{1}{2}kx^2\psi_0=E_0\psi_0$$

$$\frac{\hbar^2c^2}{2m}\psi_0-\frac{\hbar^2c^4}{2m}x^2\psi_0+\tfrac{1}{2}kx^2\psi_0=E_0\psi_0$$

$$E_0=\frac{\hbar^2c^2}{2m}-\frac{\hbar^2c^4}{2m}x^2+\tfrac{1}{2}kx^2$$

Since $c=\left(\frac{m\omega}{\hbar}\right)^{\frac{1}{2}}$; $\quad E_0=\frac{\hbar^2}{2m}\left(\frac{m\omega}{\hbar}\right)-\frac{\hbar^2}{2m}\left(\frac{km}{\hbar}\right)x^2+\tfrac{1}{2}kx^2=\tfrac{1}{2}\hbar\omega$

This is the same result as obtained from Equation 5-3:

$$E_0=(0+\tfrac{1}{2})\hbar\omega=\tfrac{1}{2}\hbar\omega\,.$$

The first several wavefunctions for the harmonic oscillator are shown in Figure 5-1 and should be compared to the Particle-in-a-Box wavefunctions shown in Figure 2-2. Note that the wavefunctions for the Particle-in-a-Box and the harmonic oscillator have similar shapes for each corresponding energy level. The principal difference is that the harmonic oscillator wavefunctions asymptotically approach zero as x approaches infinity (as the potential approaches infinity). Because the wavefunctions must

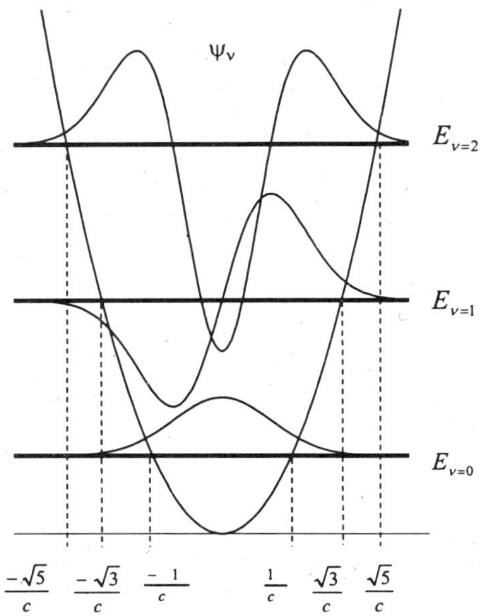

Figure 5-1. The first several harmonic oscillator wavefunctions are shown along with the classical turning points based on the quantum mechanical energy for that level.

asymptotically approach zero as x approaches infinity, this results in curvature in the wavefunctions and consequently the ground-state energy is non-zero.

To better understand the quantum mechanical harmonic oscillator, the results of the quantum mechanical system can be compared to those for the classical mechanical system (described in Section 1.3). The classical turning point for the mass, $\pm x_{max}$, occurs when the energy of a given state is equal to the maximum potential energy of the system. This is done using the ground-state quantum mechanical energy.

$$E_0 = \tfrac{1}{2}\hbar\omega = \tfrac{1}{2}kx_{max}^2 \qquad \text{(classical analogy)}$$

Solving for the classical turning points, $\pm x_{max}$:

$$x_{\max} = \pm\sqrt{\frac{\hbar\omega}{k}} = \pm\sqrt{\frac{\hbar\left(\frac{k}{m}\right)^{\frac{1}{2}}}{k}} = \pm\sqrt{\frac{\hbar}{k^{\frac{1}{2}}m^{\frac{1}{2}}}} = \pm\sqrt{\frac{\hbar}{m\omega}} = \pm\frac{1}{c}. \quad (5\text{-}11)$$

In the classical mechanical harmonic oscillator, the points $\pm x_{\max}$ correspond to where the probability of finding the particle is the greatest since the mass stops at these points (zero kinetic energy), and the probability of finding the particle beyond these points is zero.

This "leaking out" of the wavefunction as it asymptotically approaches zero can be seen quantitatively by computing the probability density of the particle beyond the classical turning points $\pm x_{\max}$. This is done specifically for the ground-state in the regions of $-\infty \le x \le -\frac{1}{c}$ and $\frac{1}{c} \le x \le \infty$ where classically it would be predicted that there should be zero probability of the particle existing because the potential energy exceeds the energy of the particle.

Probability of the Particle Beyond the classical turning points for ψ_0:

$$\int_{-\infty}^{-\frac{1}{c}} \psi_0^* \psi_0 dx + \int_{\frac{1}{c}}^{\infty} \psi_0^* \psi_0 dx = 2\int_{\frac{1}{c}}^{\infty} \psi_0^* \psi_0 dx = \frac{2c}{\sqrt{\pi}} \int_{\frac{1}{c}}^{\infty} e^{-c^2x^2} dx \approx 0.1573 \quad (5\text{-}12)$$

The integral in Equation 5-12 cannot be solved analytically; however, it can be solved numerically by setting the constant c to any value (the result is independent of the value of the constant c). As can be seen by the results in Table 5-2, the probability of the particle to exceed the classical turning points decreases as the value of v increases. This is in part because the region of space that the particle is being confined to by the potential is getting larger (note that $\pm x_{\max}$ is increasing with v), and the curvature of the function is increasing due to the increased kinetic energy of the particle.

The points of maximum and minimum amplitude for the harmonic oscillator wavefunctions (indicative of the greatest probability of the mass) can be found by taking the first derivative of the wavefunction and setting it equal to zero.

Table 5-2: The probability of the particle beyond the classical turning points, $\pm x_{max}$, for a quantum mechanical harmonic oscillator is listed below.

v	$\pm x_{max}$	**Probability Beyond the Classical Turning Points, $\pm x_{max}$**
0	$\pm\frac{1}{c}$	0.1573
1	$\frac{\sqrt{3}}{c}$	0.116
2	$\pm\frac{\sqrt{5}}{c}$	0.09507

$$\frac{d}{dx}(\psi_0)=\frac{d}{dx}\left(\sqrt{\frac{c}{\pi^{\frac{1}{2}}}}e^{-\frac{1}{2}c^2x^2}\right)=-\sqrt{\frac{c}{\pi^{\frac{1}{2}}}}c^2xe^{-\frac{1}{2}c^2x^2}=0; \qquad x=0$$

The result for the ground-state is analogous to the classical system. The greatest probability of finding the particle is at the equilibrium position for the spring. However, the ground-state energy is not zero, $E_0=\frac{1}{2}\hbar\omega$, as in the classical mechanical harmonic oscillator. The classical analogy is that the particle is not at rest even in the ground-state. The points of greatest probability densities are done for several other states, and the results are listed in Table 5-3. As the value of v increases, so does the curvature resulting in an increasing number of nodes just like in the Particle-in-a-Box wavefunctions.

Another interesting feature of the quantum mechanical harmonic oscillator is that the energy difference between subsequent levels is the same: $E_{v+1}-E_v=\hbar\omega$. This feature of uniform energy levels is a result of the symmetry of the system.

Table 5-3: The points of maximum and minimum amplitude for ψ_v for a quantum mechanical harmonic oscillator. These represent the points of maximum probability densities for the particle.

v	**Maximum and Minimum Amplitudes of** ψ_v
0	0
1	$\pm\frac{1}{c}$
2	$0, \ \pm\frac{1.58114}{c}$
3	$\pm\frac{0.6021}{c}, \ \pm\frac{2.0341}{c}$

Chemical Connection

Determine the maximum kinetic energy for a system whereby k = 533 N/m (approximate force constant for a HCl bond) and m = 1.00 kg. What are the classical turning points for this mass? Could this be measured for a macroscopic system? Repeat the computation for the case that m is the mass of a hydrogen atom, m ≈ 1.66 x 10^{-27} kg. For the mass of hydrogen, what is the energy involved in the transition from the ground-state to the first excited state? If this transition is caused by the absorption of a photon of electromagnetic radiation, what would be the wavelength of the photon and the part of electromagnetic spectrum that this photon would correspond to? The energy of the photon, E, can be related to the wavelength of the absorbed photon, λ, by using the following relationship:

$$\Delta E = E_1 - E_0 = E_{photon} = \frac{hc}{\lambda}$$

where c is the speed of light.

The system can now be expanded for describing a diatomic molecule. In this case, the spring is separated by two different masses as shown in Figure 1-2. The Schroedinger equation can be written as follows:

$$\hat{H}(x_1, x_2)\psi(x_1, x_2) = E\psi(x_1, x_2)$$

$$\left(\frac{\hat{p}_1^2}{2m_1} + \frac{\hat{p}_2^2}{2m_2} + \tfrac{1}{2}k(x_2 - x_1 - x_0)^2\right)\psi(x_1, x_2) = E\psi(x_1, x_2)$$

$$-\frac{\hbar^2}{2m_1}\frac{\partial^2\psi(x_1, x_2)}{\partial x_1^2} - \frac{\hbar^2}{2m_2}\frac{\partial^2\psi(x_1, x_2)}{\partial x_2^2}$$
$$+\tfrac{1}{2}k(x_2 - x_1 - x_0)^2\psi(x_1, x_2) = E\psi(x_1, x_2)$$

Separation of variables can be obtained by using the center-of-mass coordinate system as described in Section 1.3. Recall that the center-of-mass coordinate system introduces the following coordinates:

$$r \equiv x_2 - x_1 - x_0 \qquad\qquad s \equiv \frac{(m_1 x_1 + m_2 x_2)}{(m_1 + m_2)}$$

The coordinate r represents the displacement of the spring from its equilibrium position, and the coordinate s corresponds to the center of mass of the system. Using this coordinate system results in the following Schroedinger equation that is similar in form to the expression in Equation 1-18.

$$\left(\frac{\hat{p}_r^2}{2\mu} + \frac{\hat{p}_s^2}{2M} + \frac{1}{2}kr^2\right)\psi(r, s) = E\psi(r, s) \tag{5-13}$$

The term μ is the reduced mass, and M is the total mass of the system.

$$\mu = \frac{m_1 m_2}{m_1 + m_2} \qquad\qquad M = m_1 + m_2$$

As discussed previously in Section 1.3, the kinetic energy operator,

$$\frac{\hat{p}_s^2}{2M},$$

corresponds to the translation of the entire system in space. Since only vibrational motion is of interest and the coordinates s and r are separable, the Schroedinger equation is reduced to the coordinate r.

$$\left(\frac{\hat{p}_r^2}{2\mu}+\tfrac{1}{2}kr^2\right)\psi(r)=E\psi(r) \tag{5-14}$$

Equation 5-14 is mathematically equivalent to Equation 5-2. As a result, Equation 5-14 produces the same results as before with the reduced mass μ instead of m and the coordinate r instead of x. The following expressions are changed, and the rest of the expressions from before remain the same.

$$\omega=\sqrt{\frac{k}{\mu}} \tag{5-15}$$

$$c=\sqrt{\frac{\mu\omega}{\hbar}} \tag{5-16}$$

$$z=cr \tag{5-17}$$

Point of Further Understanding

Demonstrate that for the case that $m_1 >> m_2$, the expressions in Equations 5-15 through 5-17 result in the same expressions as in Equations 5-4, 5-6, and 5-7.

5.2 TUNNELING, TRANSMISSION, AND REFLECTION

In the case of the harmonic oscillator as discussed in the previous section, the particle has a small but not insignificant probability beyond the classical turning points. Beyond the classical turning points, the particle is penetrating into ***classically forbidden regions*** whereby the potential energy exceeds the energy of the particle. The classical analogy of this penetration is like walking through a brick wall. If the finite potential barrier is narrow, there is a probability that the particle may emerge through the barrier. This phenomenon is called ***tunneling***. Tunneling is an important topic in chemistry as it explains such phenomena as spontaneous fission reactions, transfer of electrons through insulators between two semiconductors, conformational changes of molecules, and reactions overcoming activational barriers for which the reactants have insufficient energy to overcome. The ability of a particle to tunnel has much to do with the particle's wavefunction and so the discussion here will focus on the wavefunction and how it changes with potentials.

First we will take a diversion and consider a particle that is free to travel along the x-axis (a box of infinite length). The potential is zero all along the x-axis. This means that the particle possesses only kinetic energy all along the x-axis and there are no boundary conditions. The Schroedinger equation can be readily written for this system.

$$\hat{H}(x) = -\frac{\hbar^2}{2m}\frac{d^2}{dx^2} \qquad (5\text{-}18)$$

$$-\frac{\hbar^2}{2m}\frac{d^2\psi(x)}{dx^2} = E\psi(x) \qquad (5\text{-}19)$$

The general solution to Equation 5-19 is as follows:

$$\psi(x) = Ae^{ikx} + Be^{-ikx}, \qquad k = \frac{\sqrt{2mE}}{\hbar}. \qquad (5\text{-}20)$$

Note that the general wavefunction for a free particle is the same as the general wavefunction for a Particle-in-a-Box. The coefficients A and B are determined by the boundary conditions, however, since there are no boundary conditions for the particle along the x-axis, the coefficients can take on any value. There are no points of infinite potential as in the case of a box or finite potentials that approach infinity that cause the wavefunction to asymptotically approach zero as in the harmonic oscillator. This means that the energy and momentum of the particle are not quantized. The coefficients will depend on how the system was prepared, and there are no limitations on the values of the coefficients.

The momentum of the particle can be determined by applying the momentum operator (see Equation 2-6) to the two parts of the wavefunction in Equation 5-20 independently.

$$\hat{p}(Ae^{ikx}) = \frac{\hbar}{i}\frac{d}{dx}(Ae^{ikx}) = k\hbar(Ae^{ikx}) \tag{5-21}$$

$$\hat{p}(Be^{-ikx}) = \frac{\hbar}{i}\frac{d}{dx}(Be^{-ikx}) = -k\hbar(Be^{-ikx}) \tag{5-22}$$

The only difference between the solutions in Equations 5-21 and 5-22 is the sign, so it can be concluded that since momentum is a vector quantity, the two solutions represent the particle with a momentum of equal magnitude but opposite directions. If a particle is shot from a cannon in the positive x direction, the value of the coefficient B for the wavefunction associated with that particle is zero. Likewise, a dueling particle shot in the negative x direction will have a wavefunction with a coefficient A equal to zero.

It is interesting to note that the wavefunction for the particle in this system extends from negative to positive infinity along the x-axis. Since there are no boundary conditions for the particle, there is no quantization or regions where the particle has the greatest probability density. The wavefunction is evenly distributed throughout the x-axis just like a pure wave. It can be concluded that in part what "limits" the wavelike nature of matter is the potentials that a particle experiences. In fact, part of the reason for our particulate view of matter is as a result of the potentials that all matter in the universe experiences.

Point of Further Understanding

An alternatively correct wavefunction for a free particle is

$$\psi(x) = C\cos(kx) + Di\sin(kx).$$

Is it possible for the particle to have a ground-state energy of zero? What will the real part of the wavefunction look like in the ground-state? Why must the ground-state energy immediately change when the particle becomes limited to a finite length, L, in x? How does this restriction in x result in quantization of the energy?

To explore the effect of applying a sudden potential on a particle, consider the hypothetical system shown in Figure 5-2. This system can be broken down into three separate regions. The particle originates in Region I with no potential. The energy of the particle is E, and the particle is initially moving in the positive x-direction originating in Region I. Region II begins at x = 0 and has a constant potential of V_{II}. At this point, the energy of the particle is greater than the potential in Region II. Region III extends from x = a to infinity with a zero potential. The potential changes abruptly from one region to the next.

Free particle wavefunctions for the particle in each region can be assumed since the potential is constant in each region. The general wavefunctions for each region are as follows:

$$\psi_I(x) = A_I e^{ikx} + B_I e^{-ikx}, \qquad k = \frac{\sqrt{2mE}}{\hbar}; \qquad (5\text{-}23)$$

$$\psi_{II}(x) = A_{II} e^{ik_{II}x} + B_{II} e^{-ik_{II}x}, \qquad k_{II} = \frac{\sqrt{2m(E - V_{II})}}{\hbar}; \qquad (5\text{-}24)$$

$$\psi_{III}(x) = A_{III} e^{ikx} + B_{III} e^{-ikx}, \qquad k = \frac{\sqrt{2mE}}{\hbar}. \qquad (5\text{-}25)$$

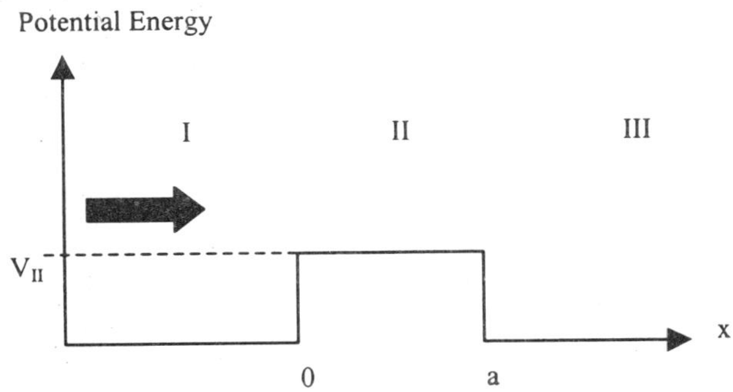

Figure 5-2. A hypothetical step potential system for a one-dimensional particle is shown. The system is divided into three different regions based on the potential energy for that region. The particle is moving in the positive x direction or to the right of diagram.

At this point, some interpretation of the coefficients for the wavefunctions of the particle in the three regions can be made. The particle is stated to be initially travelling in the positive x direction in Region I which is signified by the coefficient A_I. At this point, it cannot be ruled out that the particle may be reflected when encountering the potential in Region II even if the potential is less than the energy of the particle. The reflection of the particle back into Region I is represented by the coefficient B_I. The coefficient A_{II} represents the particle moving to the right through Region II whereas B_{II} represents the movement of the particle to the left. In Region III, the coefficient A_{III} represents the particle leaving Region II and moving through Region III to the right. The coefficient B_{III} represents the particle moving to the left in Region III. Since there is no potential in Region III, there can be no reflection in this region and the value of B_{III} must be zero. The wavefunction for Region III can be rewritten as follows:

$$\psi_{III}(x) = A_{III} e^{ikx}, \qquad k = \frac{\sqrt{2mE}}{\hbar}. \tag{5-26}$$

All three wavefunctions in Equations 5-23, 5-24, and 5-26 are associated with the particle in this system. Since there are different wavefunctions describing the particle in each of the regions, there are two important criteria that the wavefunctions must meet. Recall from Section 2.3, because the wavefunctions lead to probability densities, the wavefunctions must be single valued at the boundary of contiguous regions. It would be physically unreasonable for a particle to have two different probabilities for the same point in space. Hence, at x = 0, ψ_I and ψ_{II} must have the same value, and likewise at x = a, ψ_{II} and ψ_{III} must be equal.

At x = 0: $\psi_I(0) = \psi_{II}(0)$

$$A_I + B_I = A_{II} + B_{II} \tag{5-27}$$

At x = a: $\psi_{II}(a) = \psi_{III}(a)$

$$A_{II}e^{ik_{II}a} + B_{II}e^{-ik_{II}a} = A_{III}e^{ika} \tag{5-28}$$

The wavefunctions must also be continuous. As a result, the first derivative of ψ_I and ψ_{II} at x = 0 must be equal as well as the first derivative of ψ_{II} and ψ_{III} at x = a.

At x = 0: $\left.\dfrac{d\psi_I}{dx}\right|_{x=0} = \left.\dfrac{d\psi_{II}}{dx}\right|_{x=0}$

$$kA_I - kB_I = k_{II}A_{II} - k_{II}B_{II} \tag{5-29}$$

At x = a: $\left.\dfrac{d\psi_{II}}{dx}\right|_{x=a} = \left.\dfrac{d\psi_{III}}{dx}\right|_{x=a}$

$$k_{II}A_{II}e^{ik_{II}a} - k_{II}B_{II}e^{-ik_{II}a} = kA_{III}e^{ika} \tag{5-30}$$

Equations 5-27 through 5-30 provide four equations to obtain the four unknown coefficients (B_I, A_{II}, B_{II}, and A_{III}). The coefficient A_I represents the incoming square root of the probability density of the particle and is set by its initial conditions.

We are interested in determining the probability of the particle to be transmitted or reflected by the potential in Region II. Since at this point the energy of the particle is greater than the potential in Region II (E > V_{II}), it is classically predicted that the probability of transmission is one and that of reflection is zero. Quantum mechanically, the probability of reflection, $P^{reflection}$, is determined by the reflected probability density of the particle,

$$|B_I|^2,$$

divided by the incoming probability density of the particle,

$$|A_I|^2,$$

in Region I.

$$P^{reflection} = \frac{|B_I|^2}{|A_I|^2} \tag{5-31}$$

The transmission probability, $P^{transmission}$, is determined by the probability density in Region III,

$$|A_{III}|^2,$$

divided by the incoming probability density of the particle in Region I.

$$P^{transmission} = \frac{|A_{III}|^2}{|A_I|^2} \tag{5-32}$$

The sum of the probabilities of transmission and reflection for the particle must equal to one.

$$P^{transmission} + P^{reflection} = 1 \tag{5-33}$$

An expression for $P^{transmission}$ can be obtained by algebraically manipulating the continuity expressions in Equations 5-27 through 5-30.

$$P^{transmission} = \left(1 + \frac{\left(e^{-ik_{II}a} - e^{ik_{II}a}\right)^2}{16\left(\frac{E}{V_{II}}\right)\left(\frac{E}{V_{II}} - 1\right)} \right)^{-1} \quad (5\text{-}34)$$

Since: $2i\sin x = e^{ix} - e^{-ix}$,

$$P^{transmission} = \left(1 + \frac{\sin^2(k_{II}a)}{4\left(\frac{E}{V_{II}}\right)\left(\frac{E}{V_{II}} - 1\right)} \right)^{-1} \quad (5\text{-}35)$$

The reflection probability is readily obtained by substituting Equation 5-35 into Equation 5-33.

$$P^{reflection} = 1 - P^{transmission} = \left(1 + \frac{4\left(\frac{E}{V_{II}}\right)\left(\frac{E}{V_{II}} - 1\right)}{\sin^2(k_{II}a)} \right)^{-1} \quad (5\text{-}36)$$

The reflection probability is not necessarily zero even though the energy of the particle exceeds the potential. This phenomenon is called ***antitunneling*** or ***nonclassical scattering***.

First consider the case where the energy of the particle is *much* greater than the potential in Region II (E >> V_{II}).

$$P^{transmission} = \left(1 + \frac{\sin^2(k_{II}a)}{4\left(\frac{E}{V_{II}}\right)\left(\frac{E}{V_{II}} - 1\right)}\right)^{-1} = \left(1 + \frac{V_{II}^2 \sin^2(k_{II}a)}{4E(E - V_{II})}\right)^{-1} \cong 1$$

It is reassuring to see that the classical prediction of complete transmission of the particle is obtained at the limit that the energy of the particle is much greater than the potential in Region II.

Now consider the case where the energy is only somewhat larger than the potential in Region II ($E > V_{II}$). The sine-squared function in Equation 5-35 varies from a minimum of zero (complete transmission), to a maximum of one (a non-zero reflection probability). Complete transmission of the particle will occur when the width of Region II, a, times the constant k_{II} is equal to some positive integer, n, multiple of π.

$$\textit{Complete Transmission:} \qquad k_{II}a = n\pi \qquad (n = 1,2,3,...) \qquad (5\text{-}37)$$

The minimum transmission probability (or maximum reflection probability) occurs when the factor $k_{II}a$ is some positive odd integer, n', multiple of $\pi/2$.

$$\textit{Minimum Transmission:} \qquad k_{II}a = \frac{n'\pi}{2} \qquad (n' = 1,3,5,...) \qquad (5\text{-}38)$$

As can be seen by Equation 5-35, the transmission probability (and reflection probability) depends on the difference between the energy of the particle and the potential, mass of the particle, and on the width of the potential region.

The points in Figure 5-34a and b that represent zero reflection (or complete transmission) are called ***scattering resonances***. The large variation of transmission probabilities with incoming kinetic energy of the particle is entirely a quantum mechanical effect. Ernest Rutherford first observed this scattering phenomenon in 1909 by bombarding a thin gold foil with alpha particles. Scattering experiments continue to be the focus of many experimental and theoretical studies. These types of experiments provide much information about the interaction between particles, and it is

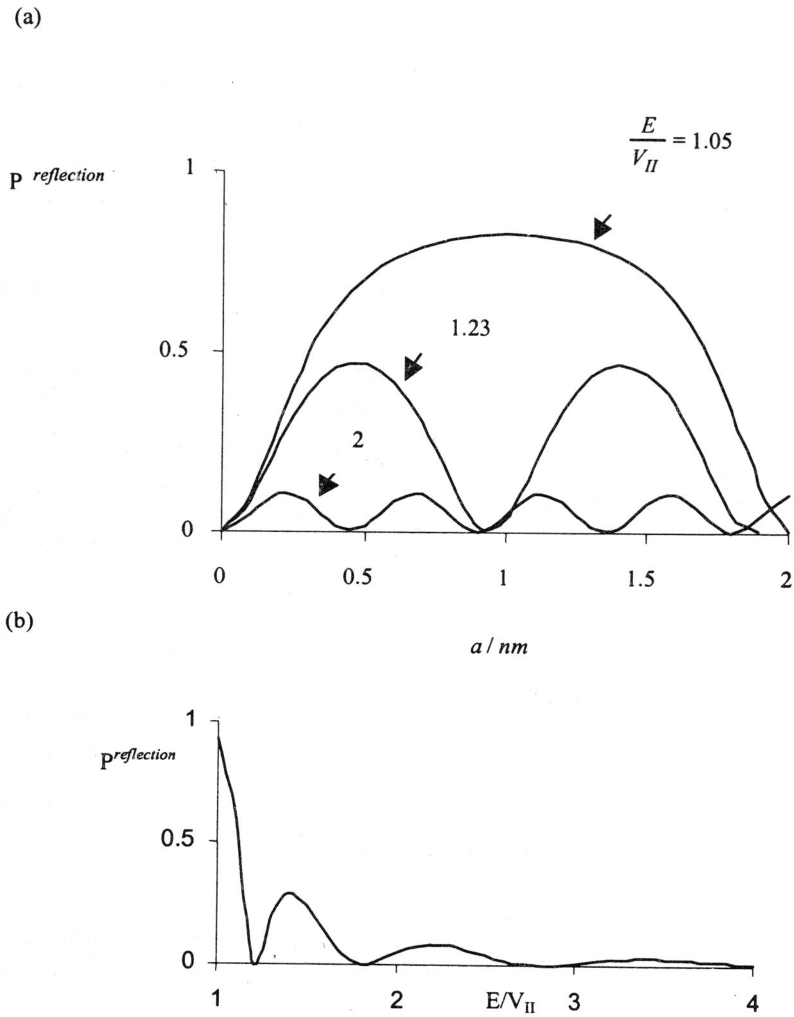

Figure 5-3. (a) The reflection probability of a proton with energy of 1 eV encountering a potential of 2 eV is shown as a function of the width the barrier. (b) The reflection probability of a proton encountering a potential of 2 eV is shown as a function of increasing kinetic energy. The width of the potential, a, is 1 nm. As the kinetic energy of the proton increases, the classical result of zero reflection is observed.

the fundamental basis for determining theoretical rate constants for chemical reactions.

Now consider the case where the potential in Region II of the hypothetical system shown in Figure 5-2 *exceeds* the energy of the particle ($V_{II} > E$). Classically the particle is forbidden to penetrate Region II; however, we have already seen in the case of the quantum mechanical harmonic oscillator that quantum mechanics predicts that the particle will have some probability of penetrating Region II and perhaps even tunneling through into Region III.

The constant k_{II} in the wavefunction for Region II will now be complex. It is convenient to write an expression for k_{II} that separates the real and imaginary parts.

$$k_{II} = i\alpha_{II} \tag{5-39a}$$

$$\alpha_{II} = \frac{\sqrt{2m(V_{II} - E)}}{\hbar} \tag{5-39b}$$

The wavefunction for Region II can be rewritten in terms of α_{II}.

$$\psi_{II}(x) = A_{II}e^{i^2\alpha_{II}x} + B_{II}e^{-i^2\alpha_{II}x} = A_{II}e^{-\alpha_{II}x} + B_{II}e^{\alpha_{II}x} \tag{5-40}$$

The wavefunction for Region II now consists of an exponentially increasing and decaying functions resulting in a non-oscillating function. *The wavefunctions of particles in classically forbidden regions do not oscillate.* The probability of transmission of the particle into Region III (tunneling) can be determined by substituting Equation 5-39a into 5-34.

$$P^{transmission} = \left(1 - \frac{\left(e^{\alpha_{II}a} - e^{-\alpha_{II}a}\right)}{16\left(\frac{E}{V_{II}}\right)\left(\frac{E}{V_{II}} - 1\right)}\right)^{-1} \tag{5-41}$$

As before, the probability of reflection is equal to one minus the probability of transmission.

The probability that the particle will tunnel to Region III depends on the mass of the particle, the width of the potential barrier a, and the difference in the energy of particle to the potential of the barrier. Figures 5-4a and 5-4b shows the tunneling probability of an electron and a proton with a kinetic energy of 1 eV (1 eV $\cong$ 1.60 x 10^{-19} J) encountering a potential barrier of 2 eV of varying width. To put the values in Figures 5-4a and 5-4b into perspective, the diameter of an atom is on the order of 0.2 nm; hence, the tunneling ability of the electron and proton are great at the atomic scale. It can also be seen in Figures 5-4a and 5-4b that the greater the mass of the particle, the less likely the particle is to tunnel into Region III. As the energy of the particle approaches the energy of the potential barrier, the tunneling probability increases as shown in Figure 5-5.

So far the potential barrier (Region II) has been of finite width. Another important type of system to consider is one in which the potential barrier is of infinite thickness such as in a particle striking armor plating. If a particle is unable to tunnel due to an infinitely thick potential barrier, there will most likely be some significant penetration at the atomic scale of the particle into the potential barrier.

To analyze the penetration ability of a particle into an infinitely thick potential barrier, a new hypothetical system is considered as shown in Figure 5-6. The system is similar to a Particle-in-a-Box, however, the potential at x = L is finite and greater than the energy of the particle. The potential at x = 0 is infinite containing the particle to the positive x-axis. The potential is zero in Region I, and Region II extends from $L \leq x \leq \infty$.

The general wavefunction for Region I will be same as in the previous systems discussed in this Section.

$$\psi_I(x) = A_I e^{ikx} + B_I e^{-ikx}, \qquad k = \frac{\sqrt{2mE}}{\hbar}$$

Since E < V_{II}, the general wavefunction for Region II will be the same as in Equation 5-40.

$$\psi_{II}(x) = A_{II} e^{-\alpha_{II} x} + B_{II} e^{\alpha_{II} x}, \qquad \alpha_{II} = \frac{\sqrt{2m(V_{II} - E)}}{\hbar}$$

a)

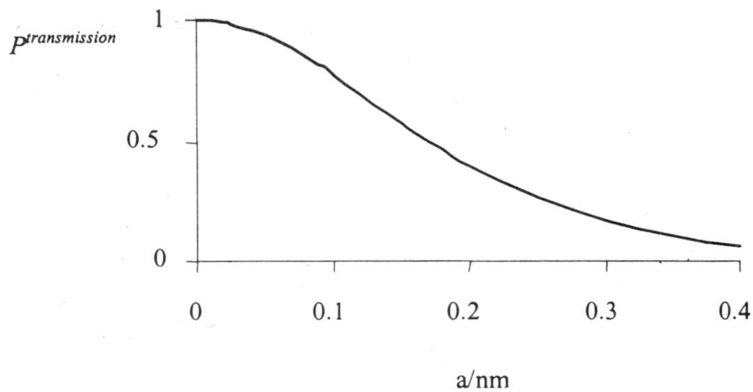

b)

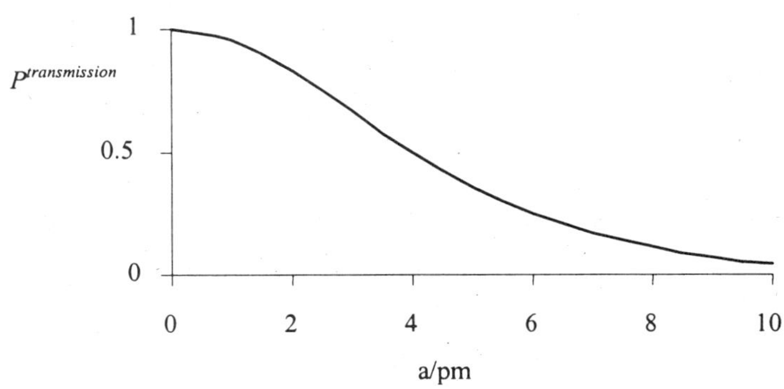

Figure 5-4. The following plots display the tunneling probabilities of an electron (a) and a proton (b) as the width of the potential barrier is varied. The electron and the proton have an energy of 1 eV, and they are both encountering a potential barrier of 2 eV.

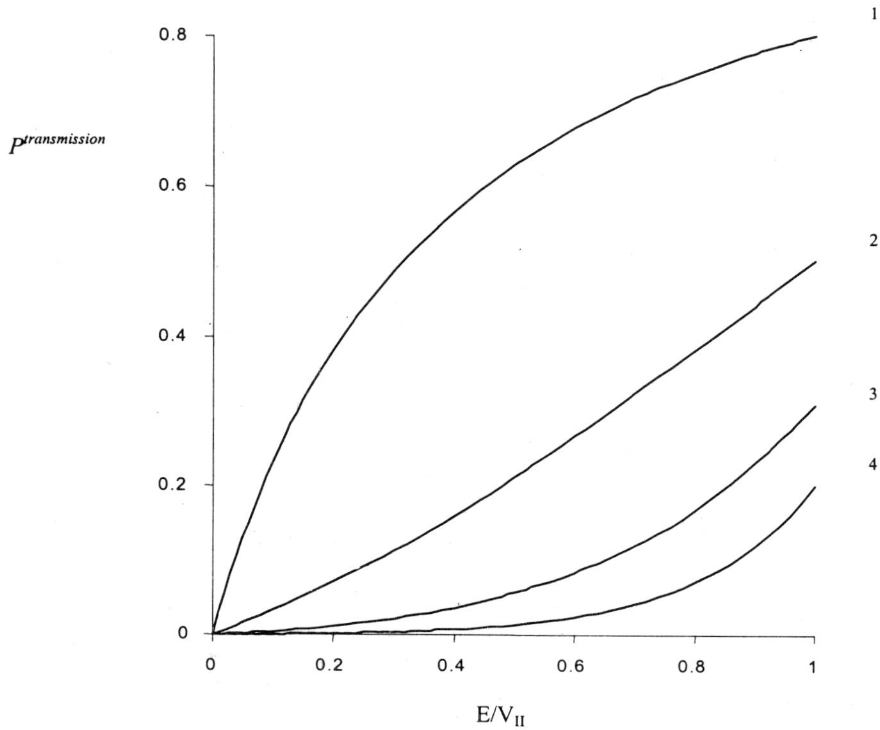

Figure 5-5. The tunneling probability of a particle as a function of its energy relative to the potential barrier is shown in this figure. Each curve represents a different value for

$$\frac{a\sqrt{2mV_{II}}}{\hbar}.$$

The wavefunction for Region II contains an exponentially increasing component as x increases. As x approaches infinity, the wavefunction in Region II will approach infinity. This is an untenable result based on the Born interpretation - the probability density of the particle will approach infinity as wavefunction approaches infinity. Since this is not physically possible, the positive exponential component of the wavefunction must be discarded.

$$\psi_{II}(x) = A_{II}e^{-\alpha_{II}x} \tag{5-42}$$

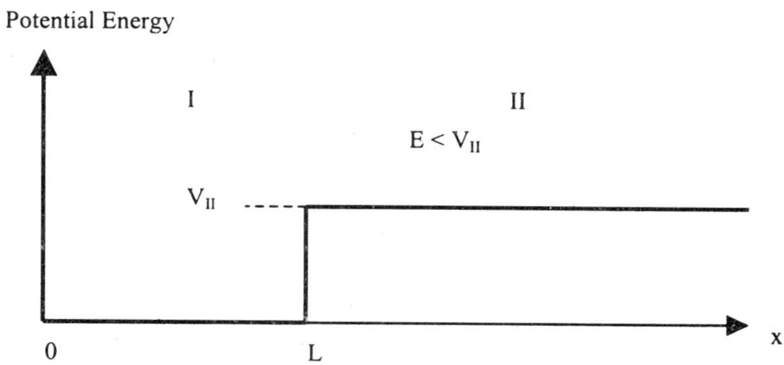

Figure 5-6. In the following system, the potential in Region I is zero except at x = 0 where the potential is infinite. The potential is constant at V_{II} throughout Region II. Region II extends from x = L to infinity. The energy of the particle, E, is less than V_{II}.

Since the potential is infinite at x = 0, the wavefunction for the particle at this point must also equal to zero.

$$\psi_I(0) = A_I + B_I = 0$$

or

$$A_I = -B_I$$

This reduces the wavefunction for the particle in Region I to the following expression:

$$\psi_I(x) = A_I\left(e^{ikx} - e^{-ikx}\right) = 2iA_I \sin(kx) = A'_I \sin(kx) . \qquad (5\text{-}43)$$

The continuity conditions can now be applied to the wavefunctions for Regions I and II at x = L. This results in the following expressions.

$$A'_I \sin(kL) = A_{III} e^{-\alpha_{II} L} \qquad (5\text{-}44)$$

and

$$kA_I' \cos(kL) = -\alpha_{II} A_{III} e^{-\alpha_{II} L}. \tag{5-45}$$

If Equation 5-44 is divided by Equation 5-45, the following expression results:

$$\tan(kL) = -\frac{k}{\alpha_{II}} = -\sqrt{\frac{E}{(V_{II} - E)}}. \tag{5-46}$$

Equation 5-46 indicates that only certain energies relative to V_{II} will result in penetration of the particle into Region II. For instance, if E << V_{II}, then tan(kL) will be nearly zero indicating essentially no penetration. Maximum penetration will occur for terms of kL that are positive integers of π/4 resulting in tan(kL) equal to one.

$$kL = \frac{n\pi}{4}, \qquad n = 1,2,3,... \tag{5-47}$$

To put the penetration ability of particles into perspective, consider an electron that is accelerated across a potential difference of 1 Volt. The electron now has a kinetic energy, E, of 1 eV (1 J/C * 1.6 x 10^{-19} C = 1.6 x 10^{-19}J). The electron approaches an infinitely wide potential barrier, V_{II}, of an opposing 2 Volts (3.2 x 10^{-19}J). The term α_{II} is computed as follows:

$$\alpha_{II} = \frac{\sqrt{2m_e(V_{II} - E)}}{\hbar}$$

$$= \frac{\sqrt{2(9.11x10^{-31}\,kg)(3.2x10^{-19} - 1.6x10^{-19})J}}{1.05x10^{-34}\,Js} = 5.14 nm^{-1}.$$

Since the energy of the particle is half of the potential energy of the barrier, α_{II} = k. To obtain the maximum penetration of the particle, Equation 5-47 is applied where n is equal to one.

$$kL = \alpha_{II} L = \frac{\pi}{4}$$

Since $\psi_{II} \propto e^{-\alpha_{II} x}$, we wish to determine at what distance (L + x) that the wavefunction has diminished, for the sake of illustration, to 0.1 of its value at x = L.

$$0.1 = e^{-5.14(L+x)} = e^{\frac{-\pi}{4}} e^{-5.14x}$$

$$x = 1.52 nm$$

The diameter of an atom is on the order of 0.2 nm indicating that the electron is capable of penetrating a depth of approximately 8 atoms. As this example portrays, the penetration of particles can have important effects on surface processes such as electrodes, heterogeneous catalysis, or any other process that occurs at the atomic scale.

PROBLEMS AND EXERCISES

5.1) Determine the wavefunction for the ν = 3 of the harmonic oscillator using (a) Equation 5-9 and (b) using Equation 5-10.

5.2) Determine the following for a harmonic oscillator in the ν = 2 state: (a) the average kinetic energy., (b) the average potential energy, (c) the average momentum, and (d) the average position of the mass. How do the quantum mechanical results compare to that for a classical system?

5.3) Calculate the probability of the particle in a harmonic oscillator to be beyond the classical turning point for the ν = 3 state.

5.4) Apply the harmonic oscillator Hamiltonian to the ψ_4 wavefunction and verify that it is an eigenfunction.

5.5) For a harmonic oscillator in the ground-state, determine Δp and Δx. How does the value of $\Delta p \Delta x$ compare to the Heisenberg Uncertainty Principle?

5.6) Locate the nodes for harmonic oscillator for the state $\nu = 7$.

5.7) The normal bond length, x_0, for an HI molecule is approximately 161 pm. The force constant of the bond is 313.8 N/m. What is the probability that the bond will be 10% greater if the molecule is in (a) the ground-state and (b) in the $\nu = 3$ excited state?

5.8) The harmonic oscillator is an approximation to describing the vibrational motion between atoms in a bond. The Morse potential is a more accurate description. The first-correction to the potential is of the following form:

$$V^{(1)} = gx^3$$

whereby g is a constant. Determine the first-order correction to the energy for a harmonic oscillator in the ground and first excited states.

5.9) Determine the maximum penetration depth of a proton that is in the same system as described in Figure 5-6. The energy of the proton is 1.6×10^{-19} J and the potential barrier is 3.2×10^{-19} J.

5.10) Make a plot similar to Figure 5-3b but instead for an electron encountering an opposing potential of 2 eV. How does mass effect the reflection probability?

Chapter 6

Vibrational/Rotational Spectroscopy of Diatomic Molecules

This chapter focuses on applying the fundamentals of quantum mechanics developed in the previous chapters to interpreting the vibrational and rotational transitions that occur within diatomic molecules in infrared spectroscopy. Analysis of an infrared spectrum of a diatomic molecule results in structural information about the molecule and the energy differences between the molecule's vibrational and rotational eigenstates.

6.1 FUNDAMENTALS OF SPECTROSCOPY

Molecular spectroscopy is a means of probing molecules and most often involves the absorption of electromagnetic radiation. The absorbed electromagnetic radiation results in transitions between eigenstates of a molecule. The type of eigenstates involved in a transition depends on the energy of the radiation absorbed. Figure 6-1 shows an electromagnetic spectrum along with the relative energies, wavelengths, and frequencies associated with each type of radiation. Absorbed ultraviolet and visible radiation generally results in transitions amongst electronic eigenstates. Absorbed infrared radiation results in changes in vibrational and rotational eigenstates. Absorbed microwave radiation results in changes in rotational eigenstates. The specific wavelengths of radiation that are absorbed in each region of the electromagnetic spectrum depend on the energy difference between the eigenstates of a molecule. As an example, a diatomic molecule with a "stiff" bond will absorb at a higher energy photon (shorter wavelength) than another diatomic molecule with a less "stiff" bond. The absorbed radiation in a spectrum provides information on the energy differences amongst various eigenstates of a molecule; however, it does not provide any information on the actual eigenstates involved in the transitions. Quantum mechanics is needed in order to analyze a spectrum in terms of assigning an absorption in a spectrum to a specific transition in eigenstates of a molecule.

The energy of a photon of electromagnetic radiation is inversely proportional to its wavelength, λ.

$$E_{photon} = \frac{hc}{\lambda} \tag{6-1}$$

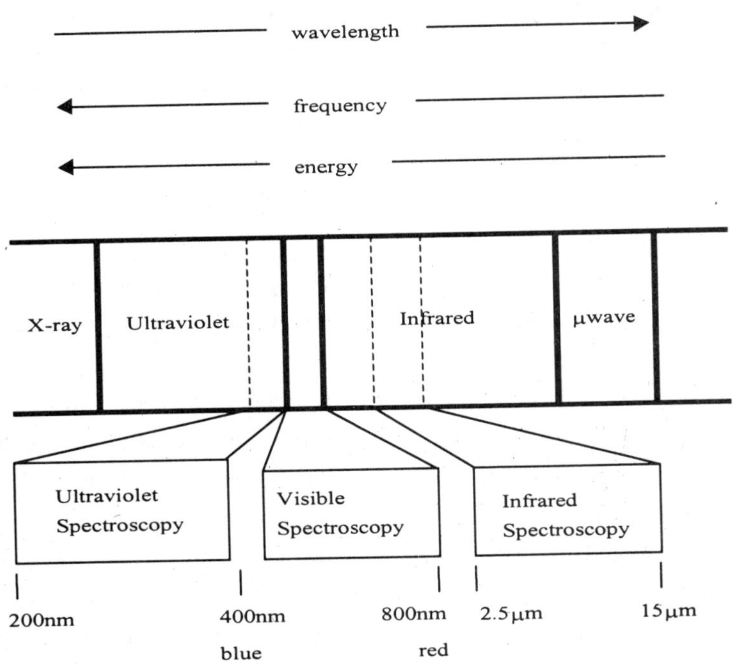

Figure 6-1. The following electromagnetic spectrum indicates the regions commonly used for ultraviolet/visible and infrared absorption spectroscopy.

The h in Equation 6-1 is Planck's constant and c is the speed of light. The wavelength of the photon absorbed in the infrared region is recorded in micrometers (microns, μm) or in terms of wavenumbers ($1/\lambda$, in cm^{-1}). In microwave spectroscopy, the spectrum is recorded in terms of the frequency, ν, of the radiation. The frequency of the radiation is determined as follows:

$$\nu = \frac{c}{\lambda}, \tag{6-2}$$

and has the units of s^{-1} or Hertz, Hz. The energy of the photon is determined by multiplying the frequency by Planck's constant, h.

$$E_{photon} = h\nu \tag{6-3}$$

The energy of the photon absorbed is equal to the energy difference between the eigenstates of a molecule.

$$\Delta E_{transition} = E_{photon} \tag{6-4}$$

6.2 RIGID ROTOR HARMONIC OSCILLATOR APPROXIMATION (RRHO)

Consider the vibration and rotation of a diatomic molecule. Since the molecule is rotating in space, the Hamiltonian is best written in terms of spherical coordinates. The potential V(r) depends only on the separation of the atoms, and it develops from the electrons and the chemical bonding that occurs between the atoms. The Schroedinger equation for a rotating and vibrating diatomic molecule is

$$\hat{H}(r,\theta,\phi)\psi(\mathrm{r},\theta,\phi) = E\psi(\mathrm{r},\theta,\phi)$$

$$-\frac{\hbar^2}{2\mu}\nabla^2\psi(\mathrm{r},\theta,\phi) + \mathrm{V(r)}\psi(\mathrm{r},\theta,\phi) = \mathrm{E}\psi(\mathrm{r},\theta,\phi)\,. \tag{6-5}$$

The term μ in Equation 6-5 represents the reduced mass of the molecule. The ∇^2 operator in spherical coordinates is given in Equation 3-11 and is substituted into Equation 6-5.

$$-\frac{\hbar^2}{2\mu r}\left(\frac{\partial^2(r\cdot\psi)}{\partial r^2}\right)-\frac{\hbar^2}{2\mu r^2}\Lambda^2\psi+V(r)\psi=E\psi \tag{6-6}$$

Recall that the operator Λ^2 (the legendrian) is only in terms of the angular coordinates (see Equation 3-12). As a result, the differential equation is separable in terms of the radial and angular parts. The wavefunction must then be a product of an angular function, Y_{lm}, and a radial function, R.

$$\psi(r,\theta,\phi)=Y_{lm}R \tag{6-7}$$

Because the radial and angular parts are separable and the molecule rotates freely in space, the angular part of Equation 6-6 is identical to the Particle-on-a-Sphere model problem developed in Section 3.2. Hence, the angular functions Y_{lm} are the spherical harmonics (Equation 3-19). The solution of the Λ^2 operator applied to Y_{lm} is known and given in Equation 3-20.

$$-\frac{\hbar^2}{2\mu r^2}\Lambda^2 Y_{lm}R=\frac{\hbar^2 l(l+1)}{2\mu r^2}Y_{lm}R \tag{6-8}$$

The result of Equation 6-8 can be substituted into Equation 6-6 resulting in a differential equation in terms of the coordinate r only.

$$-\frac{\hbar^2}{2\mu r}\left(\frac{\partial^2(r\cdot Y_{lm}R)}{\partial r^2}\right)+\frac{\hbar^2 l(l+1)}{2\mu r^2}Y_{lm}R+V(r)Y_{lm}R=EY_{lm}R \tag{6-9}$$

Dividing Equation 6-9 by Y_{lm} results in the ***two-body radial Schroedinger equation***.

$$-\frac{\hbar^2}{2\mu r}\left(\frac{\partial^2(r\cdot R)}{\partial r^2}\right)+\frac{\hbar^2 l(l+1)}{2\mu r^2}R+V(r)R=ER \tag{6-10}$$

This further demonstrates that the angular and radial components of the Schroedinger equation are separable.

It is important to note that though the angular and radial coordinates are separable, the radial function associated with the vibrational motion between the particles is also associated with the angular momentum of the molecule. Equation 6-10 represents an infinite series of differential equations for all of the possible values of l (such as $l = 0, 1, 2, 3, \ldots$). The radial function R obtained from solving Equation 6-10 will be different for each value of l and must contain that label.

It is convenient to express the radial coordinate r in terms of a new coordinate s that represents a change in the distance of separation of the atoms in the diatomic molecule from some fixed distance r_0.

$$s \equiv r - r_0 \tag{6-11}$$

The fixed distance r_0 corresponds to the point of minimum potential (the normal bond length); hence, when $r = r_0$ and $s = 0$, the potential is between the two particles is at a minimum. It is also convenient to define a new function S defined in terms of the still undetermined function R.

$$S(s) = S(r - r_0) \equiv rR(r) \tag{6-12}$$

Equation 6-10 can now be written in terms of the function S and the displacement coordinate s.

$$-\frac{\hbar^2}{2\mu r}\left(\frac{\partial^2 S}{\partial s^2}\right) + \frac{\hbar^2 l(l+1)}{2\mu(s+r_0)^2}\left(\frac{S}{r}\right) + V(s+r_0)\left(\frac{S}{r}\right) = E\left(\frac{S}{r}\right) \tag{6-13}$$

Equation 6-13 can now be multiplied by r resulting in the following expression:

$$-\frac{\hbar^2}{2\mu}\left(\frac{\partial^2 S}{\partial s^2} - \frac{l(l+1)}{(s+r_0)^2}S\right) + V(s+r_0)S = ES\,. \tag{6-14}$$

In order to solve Equation 6-14, the potential must now be specified. One possibility is that the potential is harmonic.

$$V(s+r_0)=\tfrac{1}{2}ks^2 \tag{6-15}$$

Equation 6-15 is substituted into Equation 6-14.

$$-\frac{\hbar^2}{2\mu}\left(\frac{\partial^2 S}{\partial s^2}-\frac{l(l+1)}{(s+r_0)^2}S\right)+\tfrac{1}{2}ks^2 S=ES \tag{6-16}$$

In the case that $l = 0$, Equation 6-16 is identical to the Schroedinger equation for the harmonic oscillator (see Equation 5-14). For values of l other than zero, the potential for the system is changed by the angular momentum of the system.

$$V_l^{effective}(s+r_0)\equiv V(s+r_0)+\frac{\hbar^2 l(l+1)}{2\mu(s+r_0)^2} \tag{6-17}$$

In order to solve Equation 6-17 for any value of l, the s dependence of the effective potential is expanded into a power series.

$$\frac{1}{(s+r_0)}=\frac{1}{r_0^2}-\frac{2s}{r_0^3}+\frac{6s^2}{r_0^4}-\frac{24s^3}{r_0^5}+\cdots \tag{6-18}$$

The various terms in the power series in Equation 6-18 represent the interaction of rotational momentum with vibration. Truncating the infinite series creates an expression for the effective potential. The most severe truncation is to retain only the first term. The effective potential becomes the vibrational potential plus a constant term.

$$V_l^{effective}\approx V(s+r_0)+\frac{\hbar^2 l(l+1)}{2\mu r_0^2}$$

Equation 6-16 with the severe truncation of Equation 6-18 becomes:

$$-\frac{\hbar^2}{2\mu}\left(\frac{\partial^2 S}{\partial s^2}-\frac{l(l+1)}{r_0^2}S\right)+V(s+r_0)S=ES\,. \qquad (6\text{-}19)$$

The constant operator term of the effective potential can be brought to the right side of Equation 6-19 and combined with E resulting in E'.

$$-\frac{\hbar^2}{2\mu}\frac{\partial^2 S}{\partial s^2}+V(s+r_0)S=E'S \qquad (6\text{-}20)$$

$$E'=E+\frac{\hbar^2 l(l+1)}{2\mu r_0^2} \qquad (6\text{-}21)$$

If the potential term $V(s + r_0)$ is that of the harmonic oscillator, the energy term E' is a sum of the harmonic oscillator and Particle-on-a-Sphere eigenvalues.

$$E_{vJ}=(v+\tfrac{1}{2})\hbar\omega+\frac{\hbar J(J+1)}{2\mu r_0^2} \qquad (6\text{-}22)$$

Using the designation *J* rather than *l* describes molecular rotation, and the degenerate m_l states are designated as M_J. The result in Equation 6-22 is called the ***rigid rotor harmonic oscillator (RRHO) approximation*** for a diatomic molecule.

Equation 6-22 can be further simplified by introducing a vibrational constant, ω_0, and a rotational constant, B_0.

Vibrational Constant: $$\omega_0=\hbar\omega \qquad (6\text{-}23)$$

Rotational Constant: $$B_0=\frac{\hbar^2}{2\mu r_0^2} \qquad (6\text{-}24)$$

Substitution of Equations 6-23 and 6-24 into Equation 6-22 results in the following expression for the rotational/vibrational energy of a diatomic molecule at the RRHO approximation:

$$E_{vJ} = (v + \tfrac{1}{2})\omega_0 + J(J+1)B_0 . \tag{6-25}$$

Interpreting an infrared spectrum of a diatomic molecule can assess the validity of the RRHO approximation. A Fourier transform infrared (FTIR) spectrum of hydrogen chloride ($^1H^{35}Cl$) is shown in Figure 6-2. It will be proven later in this section that when a diatomic molecule absorbs a photon of infrared radiation, the molecule will generally undergo a vibrational transition of one eigenstate ($v \rightarrow v + 1$). In addition to the change in vibrational state, the rotational state of the diatomic molecule will change by increasing or decreasing by one ($J \rightarrow J \pm 1$) based on spectroscopic selection rules to be discussed in Section 6.7. The absorption lines in an infrared spectrum correspond to a particular initial v, J eigenstate to a final v', J' eigenstate as a result of the absorption of a photon.

$$\Delta E_{photon} = E_{v'J'} - E_{vJ}$$

For the case that J increases by one (J' = J + 1) and v increases by one (v' = v + 1):

$$\Delta E_{v,J\rightarrow v+1,J+1} = [(v+1+\tfrac{1}{2})\omega_0 + (J+1)(J+1+1)B_0]_{final} \\ -[(v+\tfrac{1}{2})\omega_0 - J(J+1)B_0]_{initial}$$

$$\Delta E_{v,J\rightarrow v+1,J+1} = \omega_0 + B_0(2J+2) . \tag{6-26}$$

For the case that J decreases by one (J' = J - 1) and v increases by one (v' = v + 1):

$$\Delta E_{v.J\rightarrow v+1,J-1} = [(v+1+\tfrac{1}{2})\omega_0 + (J-1)(J-1+1)B_0]_{final} \\ -[(v+\tfrac{1}{2})\omega_0 - J(J+1)B_0]_{initial}$$

$$\Delta E_{v,J\rightarrow v+1,J-1} = \omega_0 - B_0 2J . \tag{6-27}$$

Equations 6-26 and 6-27 represent the predictions for the absorption lines for an infrared spectrum of a diatomic molecule based on the RRHO approximation.

Table 6-1. The predicted position, ΔE, and separation of peaks, Δ(ΔE), of a diatomic molecule in an infrared spectrum using the RRHO approximation.

J(initial) →	J'(final)	ΔE	Δ(ΔE)
4	3	$\omega_0 - 8B_0$	
			$2B_0$
3	2	$\omega_0 - 6B_0$	
			$2B_0$
2	1	$\omega_0 - 4B_0$	
			$2B_0$
1	0	$\omega_0 - 2B_0$	
			$4B_0$
0	1	$\omega_0 + 2B_0$	
			$2B_0$
1	2	$\omega_0 + 4B_0$	
			$2B_0$
2	3	$\omega_0 + 6B_0$	
			$2B_0$
3	4	$\omega_0 + 8B_0$	
			$2B_0$
4	5	$\omega_0 + 10B_0$	

The predicted spectral lines of a diatomic molecule, ΔE, and the distance of separation between successive spectral lines, Δ(ΔE), for various transitions in the rotational state J can now be determined. This is shown in Table 6-1.

The RRHO approximation predicts that the infrared spectrum of a diatomic molecule will have a number of peaks all of equal separation ($2B_0$) except one larger gap of $4B_0$ between the set of peaks where J is increasing by one (R-branch) and the set where J is decreasing by one (P-branch). As can be seen by Figure 6-2, the RRHO approximation predicts the infrared spectrum of a diatomic molecule remarkably well in spite of the harmonic oscillator approximation for vibrational motion and the severe truncation of the power series in Equation 6-18. The distinct gap between the P and R-branches can be clearly seen. However, note that the distance of separation between the peaks in the infrared spectrum has some variation whereas the

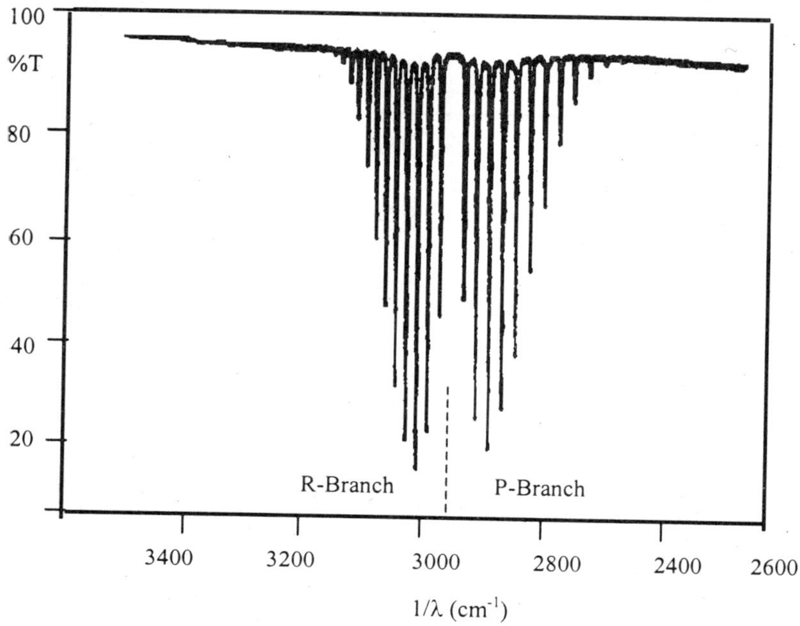

Figure 6-2. An FTIR spectrum of hydrogen chloride ($^1H^{35}Cl$) at room temperature (298 K) is shown. The general pattern of absorption peaks is predicted by the RRHO approximation; however, increasing deviation from the RRHO approximation occurs as the initial J state increases.

RRHO approximation predicts it to be constant. The RRHO approximation does not take into account that the rotational motion cannot be entirely separated from the vibrational motion of the molecule, and the vibrational motion is not strictly harmonic oscillations. This will be taken into account in the following sections by including additional terms in the series expansion of $V_l^{effective}$ (Equation 6-18) and by correcting for some anharmonicity in the vibrational motion.

The intensity of the peaks and the actual vibrational transitions ($v = 0 \rightarrow v' = 1$; or $v = 1 \rightarrow v' = 2$; etc.) that are observed in the spectrum depends primarily on the number of molecules in the initial eigenstate before the absorption of infrared radiation. If the molecules of the system are in thermal equilibrium at a temperature T, the fraction of molecules, f_i, in a given quantum state E_i is described by the ***Maxwell-Boltzmann distribution law.***

The Maxwell-Boltzmann distribution law is simply stated here, but it is developed and proven in a number of texts, some of which are listed in the bibliography section at the end of this chapter. The fraction of molecules, f_i, in a given state E_i is

$$f_i = \frac{g_i e^{-\frac{E_i}{kT}}}{q}. \tag{6-28}$$

The term k is the Boltzmann constant (approximately 1.38 x 10^{-23} J K^{-1}), q is the ***molecular partition function***, and g_i is the degeneracy of the E_i eigenstate. The molecular partition function is a sum over all the quantum states for a molecule (i.e. an infinite sum).

$$q = \sum_j g_j e^{-\frac{E_j}{kT}} \tag{6-29}$$

Since the fraction of molecules in an eigenstate E_i depends on the average thermal energy of a molecule, kT, and the ground-state is the lowest possible eigenstate, the ground-state energy is taken as zero in the Maxwell-Boltzmann distribution.

The fraction of molecules in each vibrational state can now be determined. The energy eigenstates for the harmonic oscillator setting the ground-state energy as zero is

$$E_v = (v + \tfrac{1}{2} - \tfrac{1}{2})\hbar\omega = v\omega_0. \qquad (v = 0,1,2,\ldots) \tag{6-30}$$

Equation 6-30 can now be substituted into Equation 6-29 to obtain the vibrational molecular partition function.

$$\begin{aligned} q_{vib} &= \sum_v e^{-\frac{E_v}{kT}} = 1 + e^{-\frac{\omega_0}{kT}} + e^{-\frac{2\omega_0}{kT}} + e^{-\frac{3\omega_0}{kT}} + \cdots \\ &= 1 + x + x^2 + x^3 + \cdots \\ &\quad \text{where } x = e^{-\frac{\omega_0}{kT}} \end{aligned} \tag{6-31}$$

Equation 6-31 is a geometric series for which the solution is well known.

$$q_{vib} = 1 + x + x^2 + x^3 + \cdots = \frac{1}{1-x} = \frac{1}{1 - e^{-\frac{\omega_0}{kT}}} \qquad (6\text{-}32)$$

Equation 6-32 can now be substituted into Equation 6-28 resulting in an expression for the fraction of molecules in a vibrational state ν at a temperature T.

$$f_v = \frac{e^{-\frac{v\omega_0}{kT}}}{q_{vib}} = e^{-\frac{v\omega_0}{kT}}\left(1 - e^{-\frac{\omega_0}{kT}}\right) \qquad (6\text{-}33)$$

Example 6-1

Problem: Determine the fraction of $^{1}H^{35}Cl$ molecules in the ground-state (ν = 0) at room temperature (298 K), and the most probable vibrational transition observed in Figure 6-2.

Solution: The vibrational constant, ω_0, for $^{1}H^{35}Cl$ is given in Table 6-2 (found in Section 6.6) as 2989.7 cm^{-1} (5.9465×10^{-20} J). The vibrational constant can also be estimated from the spectrum in Figure 6-2. The vibrational constant is substituted into Equation 6-33. Since vibrational states are non-degenerate, $g_v = 1$.

$$f_{v=0} = (1)\left(1 - e^{-\frac{5.9465 x 10^{-20} J}{(1.38 x 10^{-23} JK^{-1})(298 K)}}\right) \cong 1$$

The result indicates that essentially 100% of the $^{1}H^{35}Cl$ molecules are in the ground vibrational state. As a result, the only vibrational transition that will be observed in an infrared spectrum of $^{1}H^{35}Cl$ at room temperature is $v = 0 \rightarrow v' = 1$.

As can be seen in Example 6-1 for $^{1}H^{35}Cl$, most diatomic molecules are in the ground vibrational state at room temperature. The energy difference between vibrational levels of a diatomic molecule are so large that only at very high temperatures will there be a significant probability that a diatomic molecule will be in an excited vibrational state.

> ***Point of Further Understanding***
> Using the FTIR spectrum in Figure 6-2 and the RRHO approximation, estimate the vibrational and rotational constants of $^{1}H^{35}Cl$. Since the RRHO approximation fails to account for vibration and rotation interaction that becomes increasingly important at higher rotational states, which peaks are best to use to obtain these constants? Based on the range of infrared radiation generally used in an infrared spectrum (as shown Figure 6-1), why is not possible to see a vibrational transition over two eigenstates for $^{1}H^{35}Cl$?

Since diatomic molecules are all essentially in the ground vibrational state at room temperature, the difference in the peak heights in the infrared spectrum is primarily due to the populations of molecules in the different rotational states J. Each J rotational state is degenerate due to the 2J + 1 possible M_J states. The fraction of molecules in a given rotational state J is

$$f_J = \frac{(2J+1)e^{-\frac{J(J+1)B_0}{kT}}}{q_{rot}}. \tag{6-34}$$

The relative heights of the peaks in an infrared spectrum of a diatomic molecule can be related to the ratio of the fraction of molecules in a rotational state J, M_J compared to the fraction of molecules in the ground rotational state ($J = 0$, $M_J = 0$).

$$\frac{f_J}{f_0} = (2J+1)e^{-\frac{J(J+1)B_0}{kT}} \tag{6-35}$$

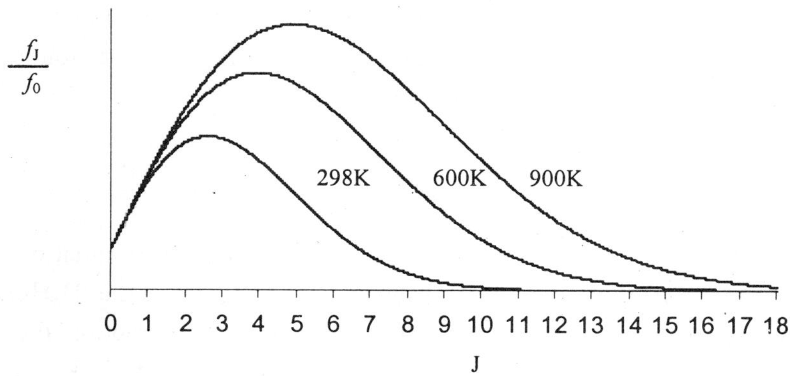

Figure 6-3. A plot of the relative population of a rotational state J to the J = 0 state $^1H^{35}Cl$ at different temperatures. The rotational constant, B_0, for $^1H^{35}Cl$ is 10.59 cm^{-1} (see Table 6-2). The curves are a plot of Equation 6-35 with J treated as a continuous variable. The relative transition intensities in an infrared spectrum of a diatomic molecule can be related to the relative populations of the rotational states.

A plot of the function in Equation 6-35 for $^1H^{35}Cl$ is shown in Figure 6-3. At room temperature (298 K), the rotational states with the highest population are at J = 3. The peaks with the greatest intensity in an infrared spectrum for $^1H^{35}Cl$ will correspond to the change in rotational states of 3 $\rightarrow$ 4 and 3 $\rightarrow$ 2. This is confirmed by the spectrum in Figure 6-2. In addition, as the temperature increases, the number of peaks that will appear in the P and R branches of the infrared spectrum increases, and the peaks of greatest intensity will correspond to increasingly larger initial J states.

The RRHO approximation and analysis of the infrared spectrum of $^1H^{35}Cl$ formulates a picture of the vibration-rotation energy levels of a diatomic molecule. The energy difference between vibrational energy levels is large with respect to the rotational energy levels. A vibrational state v will have an infinite manifold of J rotational states. This is depicted in Figure 6-4.

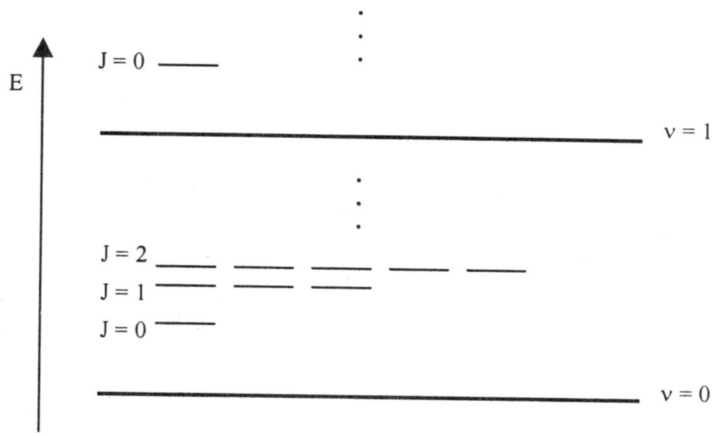

Figure 6-4. The vibrational-rotational energy eigenstates for a diatomic molecule are shown. The ground-state corresponds to v = 0, J = 0.

Point of Further Understanding

The molecular partition function, q, indicates the number of eigenstates that are accessible at a temperature. The vibrational partition function, q_{vib}, for $^{1}H^{35}Cl$ at 298K is equal to one (see Example 6-1) indicating that only the one vibrational eigenstate, the ground-state, is accessible. The rotational partition function, q_{rot}, can be estimated by replacing the summation in Equation 6-29 with an integration. This approximation is justified since the energy difference between the rotational eigenstates of $^{1}H^{35}Cl$ are close together and many states are occupied (the energy difference between rotational eigenstates gets smaller as J increases). The rotational partition function for a heteronuclear diatomic molecule with a small rotational constant such as $^{1}H^{35}Cl$ is as follows:

$$q_{rot} = \sum_{J} g_{J} e^{-\frac{E_J}{kT}} = \sum_{J} (2J+1) e^{-\frac{B_0 J(J+1)}{kT}} \cong \int_{0}^{\infty} (2J+1) e^{-\frac{B_0 J(J+1)}{kT}} dJ = \frac{kT}{B_0}$$

The molecular partition function is dimensionless; hence, the rotational constant must be converted into energy units. Calculate the rotational partition function for $^1H^{35}Cl$ at 298 K. Does your answer agree with the number of accessible rotational states J at this temperature predicted in Figure 6-3 and the number observed in the infrared spectrum in Figure 6-2? *Hint:* Remember that the degeneracy of each rotational eigenstate is 2J + 1.

6.3 VIBRATIONAL ANHARMONICITY

The realistic vibrational potentials of molecules are not strictly harmonic oscillations. The energy differences between vibrational levels are not uniform as predicted by the harmonic oscillator model problem but rather continuously decrease and form a continuum at sufficiently large vibrational eigenstates. In addition, all molecules will dissociate if promoted to a sufficiently high vibrational eigenstate. Vibrational anharmonicity refers to those parts of the stretching potential that are not harmonic, in other words, the parts of the potential that do not vary as the square of the displacement.

An approximate approach for modeling the anharmonicity of the stretching potential of a diatomic molecule is the ***Morse potential***. The Morse potential is constructed such that D_e is the depth of the minimum of the curve (related to the dissociation energy of the diatomic molecule) and choosing a parameter γ that yields the correct shape of the potential curve.

$$V(s) = D_e(1 - e^{-\gamma s})^2 \qquad (6\text{-}36)$$

The coordinate s is the displacement of the bond from its equilibrium position r_0 as defined in Equation 6-11. The qualitative form of the Morse potential is shown in Figure 6-5. At s = 0, the potential is zero. As s approaches infinity, the value of the potential approaches D_e indicating dissociation of the bond. As a result, the Morse potential has a finite number of states ($\nu = 0, 1, 2, 3 \ldots \nu_{max}$) such that when $\nu = \nu_{max}$, the bond is at the highest possible vibrational state before dissociation. For s < 0, representing compression of the bond, the Morse potential rises very rapidly as in a real molecule. The Morse potential can be compared to the harmonic oscillator potential by writing Equation 6-36 in an infinite power series in terms of s.

$$V(s) = D_e\left(\gamma^2 s^2 - \gamma^3 s^3 + \tfrac{7}{12}\gamma^4 s^4 - \tfrac{1}{4}\gamma^5 s^5 + \cdots\right) \qquad (6\text{-}37)$$

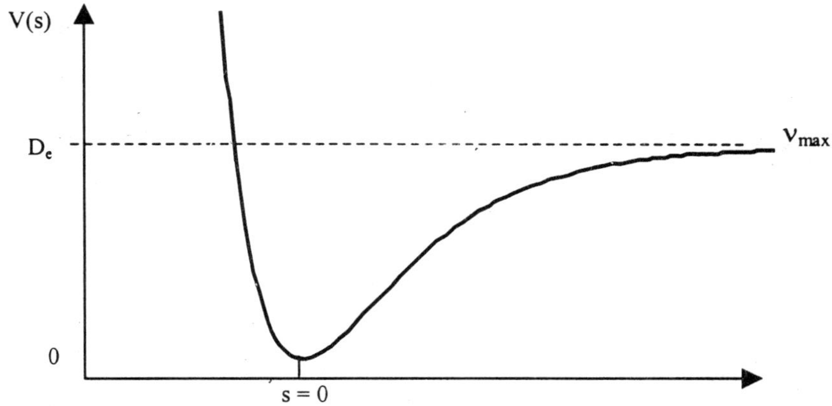

Figure 6-5. The functional form of the Morse potential (Equation 6-36) is shown. Bond dissociation occurs at $V(\infty) = D_e$. For small values of s (lowest vibrational states) the function is parabolic. Note that the potential increases rapidly for $s < 0$.

The first term in Equation 6-37 is harmonic and the subsequent third, fourth, and higher order terms are varying degrees of anharmonicity.

In infrared spectroscopy of diatomic molecules, the vibrational motion is generally limited to the first two vibrational states of a diatomic molecule whereby the displacement of the bond is near the minimum (i.e. small values of s). As a result, it is reasonable as a first approximation to confine the anharmonicity to the third order term of Equation 6-37. The potential can be represented by a third order polynomial such that the first term is the same as in the harmonic oscillator model problem.

$$V(s) = \tfrac{1}{2} ks^2 + gs^3 \tag{6-38}$$

where $\gamma = \sqrt{\dfrac{k}{2D_e}}$ and $g = -D_e\gamma^3$

Perturbation theory can now be used to determine the correction to the energy eigenvalues for vibrational motion. The first-order perturbing Hamiltonian is the cubic term in Equation 6-38.

$$\hat{H}^{(1)}(s) = gs^3 \tag{6-39}$$

The first-order correction to the energy is calculated by using Equation 4-13.

$$E_v^{(1)} = \left\langle \psi_v^{(0)} \middle| gs^3 \middle| \psi_v^{(0)} \right\rangle = 0$$

The first-order correction to the energy for all states is zero. The second-order corrections to the energy due to the gs^3 term are computed using Equation 4-18b.

$$E_v^{(2)} = \frac{\left\langle \psi_v^{(0)} \middle| gs^3 \middle| \psi_{v+1}^{(0)} \right\rangle^2}{E_v^{(0)} - E_{v+1}^{(0)}} + \frac{\left\langle \psi_v^{(0)} \middle| gs^3 \middle| \psi_{v+2}^{(0)} \right\rangle^2}{E_v^{(0)} - E_{v+2}^{(0)}} + \cdots + \frac{\left\langle \psi_v^{(0)} \middle| gs^3 \middle| \psi_{v_{\max}}^{(0)} \right\rangle^2}{E_v^{(0)} - E_{v_{\max}}^{(0)}}$$

The general solution to the second order correction to the energy is as follows:

$$E_v^{(2)} = -\frac{7}{16}\frac{g^2\hbar^2}{\mu k^2} - \frac{15}{4}\frac{g^2\hbar^2}{\mu k^2}\left(v + \tfrac{1}{2}\right)^2 \tag{6-40a}$$

or

$$E_v^{(2)} = -\frac{7}{128}\left(\frac{\omega_0^2}{D_e}\right) - \frac{15}{32}\left(\frac{\omega_0^2}{D_e}\right)\left(v + \tfrac{1}{2}\right)^2. \tag{6-40b}$$

As can be seen in Equation 6-40b, the effect of the anharmonic term gs^3 is to lower the energy for each vibrational energy eigenstate relative to the harmonic oscillator. This is shown in Figure 6-6 for $^1H^{35}Cl$. The energy levels for an anharmonic oscillator become more closely spaced with increasing energy.

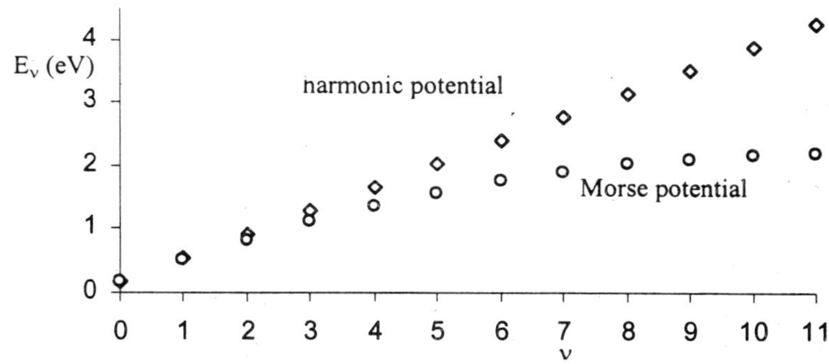

Figure 6-6. Comparison of the vibarational energy eigenstates for $^{1}H^{35}Cl$ computed using the harmonic oscillator approximation and the gs^3 term from the Morse potential (Equation 6-40b). The value of D_e is 4.80 eV obtained from Table 6-2.

The energy of a vibrational eigenstate using the cubic term of the Morse potential to a second-order energy correction is as follows:

$$E_\nu \cong E_\nu^{(0)} + E_\nu^{(1)} + E_\nu^{(2)}$$
$$= (\nu + \tfrac{1}{2})\omega_e - \frac{7}{128}\left(\frac{\omega_0^2}{D_e}\right) - \frac{15}{32}\left(\frac{\omega_0^2}{D_e}\right)(\nu + \tfrac{1}{2})^2. \quad (6\text{-}41)$$

For a change in vibrational states of $\nu = 0 \rightarrow \nu' = 1$ (as in a typical infrared spectrum of a diatomic molecule), Equation 6-41 becomes:

$$\Delta E = E_1 - E_0 \cong \omega_0 - \frac{30}{32}\left(\frac{\omega_0^2}{D_e}\right). \quad (6\text{-}42)$$

Equation 6-42 suggests that the ***dissociation energy, D_0***, of a diatomic molecule can be estimated from an infrared spectrum. The dissociation energy of a diatomic molecule, D_0, is defined as the difference between the depth of the potential well, D_e, and the ground-state energy, E_0.

$$D_0 \equiv D_e - E_0 \tag{6-43}$$

The Morse potential is only an approximation to an actual molecular oscillation. A more accurate representation is obtained via a polynomial that can be fitted to experimental data.

$$E_v = (v + \tfrac{1}{2})\omega_0 + (v + \tfrac{1}{2})^2 \omega_0 \chi_0 + (v + \tfrac{1}{2})^3 \omega_0 y_0 + \cdots \tag{6-44}$$

The term $\omega_0\chi_0$ is the first (equilibrium) anharmonicity constant, $\omega_0 y_0$ is the second anharmonicity constant, and so on.

6.4 CENTRIFUGAL DISTORTION

The energy eigenvalue expression for the vibration-rotation of a diatomic molecule can be improved by including more terms from Equation 6-18. If one additional term is added, the approximation becomes:

$$\frac{1}{(s + r_0)} \cong \frac{1}{r_0} - \frac{2s}{r_0^3}.$$

This makes the effective potential from Equation 6-17 as follows:

$$V_J^{eff}(s + r_0) \cong V(s + r_0) + a + bs$$

where $a = B_0 J(J+1)$ and $b = -\dfrac{2B_0 J(J+1)}{r_0}.$

If the vibrational potential is assumed to be harmonic, the effective potential becomes the following quadratic polynomial:

$$V_J^{eff} = \tfrac{1}{2} ks^2 + bs + a. \tag{6-45}$$

The two-body vibrating/rotating Schroedinger equation becomes:

$$\left[-\frac{\hbar^2}{2\mu}\frac{d^2\psi_v}{ds^2}+(\tfrac{1}{2}ks^2+a)\psi_v\right]+bs\psi_v=E\psi_v. \qquad (6\text{-}46)$$

One method for solving Equation 6-46 is to use perturbation theory. The term in the brackets in Equation 6-46 can be recognized as the Schroedinger equation for the RRHO approximation (Equation 6-19), and the term *bs* can be taken as a first-order perturbation. The first-order and higher order corrections to the energy eigenvalues to the RRHO approximation can then be computed using Perturbation Theory.

An easier approach to solving Equation 6-46 is recognize that the effective potential in Equation 6-45 is still a quadratic equation resulting in a parabola. As can be seen in Figure 6-7, the only difference between parabola $\frac{1}{2}ks^2$ and the parabola from the effective potential in Equation 6-45 is that the minimum potential is no longer at s = 0. The minimum is shifted by amount δ, and the minimum potential is now V_0.

$$V_J^{eff}(\delta+r_0)=V_0$$

Since the minimum of a parabolic potential of a harmonic oscillator is shifted, the effect is to add to each energy eigenvalue. As a result, each energy eigenvalue has a term V_0 added to it.

$$-\frac{\hbar^2}{2\mu}\frac{d^2\psi_v}{ds^2}+(V_0+\tfrac{1}{2}ks^2)\psi_v=[V_0+(v+\tfrac{1}{2})\omega_e]\psi_v \qquad (6\text{-}47)$$

The wavefunctions ψ_v that satisfy Equation 6-47 are the same as for a harmonic oscillator since the Schroedinger equation has the same functional form. Since the value of V_0 represents the minimum of V^{eff} at s = δ, the value of δ is determined by taking the derivative of V^{eff} (Equation 6-45) and setting the derivative equal to zero.

$$\frac{dV^{eff}}{ds}=ks+b$$

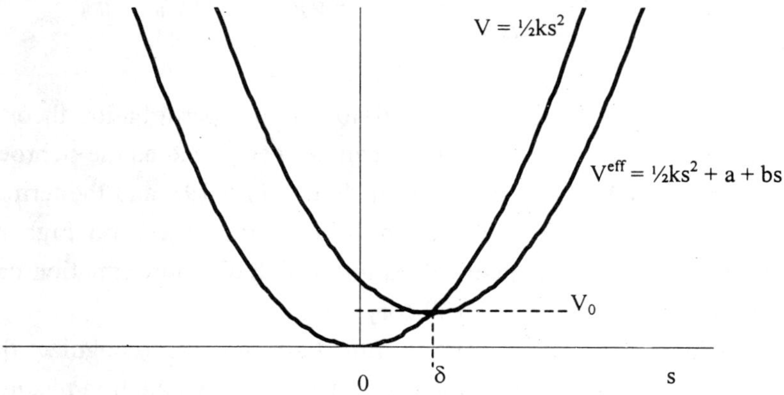

Figure 6-7. A plot of the functions $V(s) = \frac{1}{2}ks^2$ and $V^{eff}(s) = \frac{1}{2}ks^2 + a + bs$ are shown. The difference between these two functions is that the minimum for V^{eff} is at $s = \delta$ and the value of its minimum is V_0 rather than zero.

$$k\delta + b = 0$$

$$\delta = -\frac{b}{k} = \frac{2B_0 J(J+1)}{kr_0} \qquad (6\text{-}48)$$

The value of V_0 is determined by substituting Equation 6-48 into Equation 6-45.

$$V_0 = \tfrac{1}{2}k\left(\frac{-b}{k}\right)^2 + a + b\left(\frac{-b}{k}\right) = -\tfrac{1}{2}\left(\frac{b^2}{k}\right) + a$$

or

$$V_0 = -\frac{B_0^2[J(J+1)^2]^2}{2kr_0^2} + B_0 J(J+1) \qquad (6\text{-}49)$$

The energy eigenvalues are as follows:

$$E_{vJ} = (v + \tfrac{1}{2})\omega_0 + B_0 J(J+1) - D_0[J(J+1)]^2 \qquad (6\text{-}50)$$

where $$D_0 = \frac{B_o^2}{kr_0^2}. \qquad (6\text{-}51)$$

The energy eigenvalue expression in Equation 6-50 is the same as for the RRHO approximation (Equation 6-25) except for the last term. The term D_0 is called the ***centrifugal distortion constant***. Note that it does have the same symbol as dissociation energy; however, its context will indicate whether it represents dissociation energy or the centrifugal distortion constant.

The physical interpretation of centrifugal distortion is that as a result of the rotation of the diatomic molecule, the "spring" representing the bond between the atoms is stretched. This increases the effective equilibrium bond length of the molecule lowering the energy of each eigenstate. The effect of centrifugal distortion increases with increasing value of J.

6.5 VIBRATION-ROTATION COUPLING

Including an additional term can make a further improvement to the truncation made in Equation 6-18.

$$\frac{1}{(s + r_0)} \cong \frac{1}{r_0^2} - \frac{2s}{r_0^3} + \frac{6s^2}{r_0^4} \qquad (6\text{-}52)$$

The effective potential becomes as follows:

$$V_J^{eff} = V(s + r_e) + J(J+1)B_0\left(1 - \frac{2s}{r_0} + \frac{6s^2}{r_0^2}\right). \qquad (6\text{-}53)$$

If the original vibrational potential, $V(s + r_e)$, is taken to be harmonic, the effective potential has the same form as in Equation 6-45 with a different effective force constant:

$$V_J^{eff}(s + r_0) = \tfrac{1}{2}(k + 2c)s^2 + bs + a \qquad (6\text{-}54)$$

$$\text{where} \qquad c = \frac{6B_e J(J+1)}{r_0^2}. \tag{6-55}$$

The vibrational frequency will now depend on the rotational state J.

$$\omega_J = \sqrt{\frac{(k+2c)}{\mu}} = \sqrt{\frac{k}{\mu} + \frac{12B_0 J(J+1)}{\mu r_0^2}} \tag{6-56}$$

The effect of the additional term *2c* is to increase the effective vibrational frequency, ω_J, causing an increase in energy for each eigenstate except for when J = 0. This effect is called ***vibration-rotation coupling***.

Another more standard approach is to treat the additional term in the effective potential as a perturbation. Perturbation theory yields the following vibration/rotation energy eigenvalues:

$$\begin{aligned} E_{vJ} &= (v + \tfrac{1}{2})\omega_0 + J(J+1)B_0 \\ &- D_0[J(J+1)]^2 + \alpha_0(v + \tfrac{1}{2})J(J+1) \end{aligned} \tag{6-57}$$

$$\text{where} \qquad \alpha_0 = \frac{3\hbar B_0}{\omega_0 \mu r_0^2}. \tag{6-58}$$

The third term is the centrifugal distortion term, and the constant α is called the ***vibration-rotation coupling constant***. Note that the vibration-rotation coupling term involves both the vibrational and rotational quantum states v and J.

6.6 SPECTROSOPIC CONSTANTS FROM VIBRATIONAL SPECTRA

The vibrational/rotational energy states of a diatomic molecule can now be written to include not only the RRHO approximation but also in terms of the correction factors including the first anharmonicity correction, centrifugal distortion, and vibration-rotation coupling (Section 6.3 – 6.5).

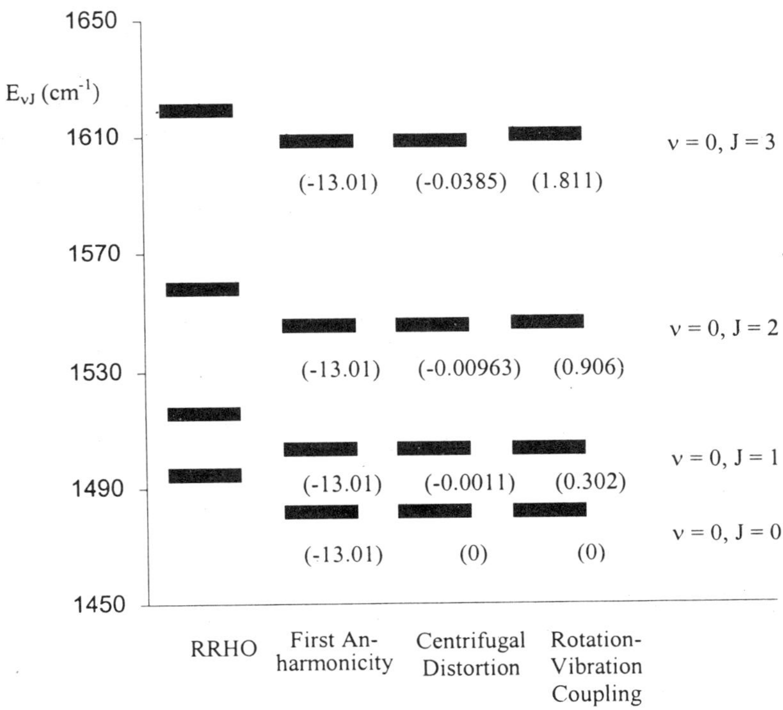

Figure 6-8. The contribution of the first anharmonicity, centrifugal distortion, and rotation-vibration coupling for $^{1}H^{35}Cl$ vibration/rotation energy levels relative to the energy values computed from the rigid rotor harmonic oscillator approximation (RRHO). The numbers in parenthesis correspond to the contribution of each correction term. The constants were obtained from Table 6-2.

$$E_{vJ} = (v + \tfrac{1}{2})\omega_0 + J(J+1)B_0 + \omega_0\chi_0(v + \tfrac{1}{2})^2 + D_0[J(J+1)]^2 + \alpha_0(v + \tfrac{1}{2})J(J+1) \tag{6-59}$$

The effect and the order of magnitude of each of the correction terms relative to the RRHO approximation for $^{1}H^{35}Cl$ are shown Figure 6-8.

The form of Equation 6-59 is still only an approximation; however, it can be generalized by realizing that the dependence on v is always in terms of v

+ ½ and the J dependence is always in terms of J(J + 1). Equation 6-59 is a truncation of the following general polynomial:

$$E_{vJ} = \sum_{i=0}^{\infty}\sum_{k=0}^{\infty} c_{ik}\,(v+\tfrac{1}{2})^{i}\,[J(J+1)]^{k}\ . \qquad (6\text{-}60)$$

The constants c_{ik} that are generated by the expansion of Equation 6-60 can be assigned to the various physical constants such as the first anharmonicity, centrifugal distortion, vibration-rotation coupling, and so on.

$$\begin{aligned} E_{vJ} &= c_{00} + \{c_{10}(v+\tfrac{1}{2}) + c_{20}(v+\tfrac{1}{2})^2 + c_{30}(v+\tfrac{1}{2})^3 + \cdots\} \\ &\quad + \{c_{01}J(J+1) + c_{02}[J(J+1)]^2 + \cdots\} \\ &\quad + \{c_{11}(v+\tfrac{1}{2})J(J+1) + \cdots\} + \cdots \end{aligned} \qquad (6\text{-}61)$$

The values of the reported coefficients in Equation 6-61 will depend on the level of truncation. The coefficients can be readily correlated to specific spectroscopic constants. Due to convention, some of the spectroscopic constants correspond to the negative of the coefficient.

ω_0	vibrational constant (c_{10})
B_0	rotational constant (c_{01})
$\omega_0\chi_0$	first-anharmonicity constant ($-c_{20}$)
$\omega_0 y_0$	second-anharmoncity constant (c_{30})
D_0	centrifugal distortion constant ($-c_{02}$)
α_0	vibration-rotation coupling (c_{11})

Table 6-2 lists spectroscopic constants for a number of diatomic molecules. The values are given in wavenumbers (cm^{-1}) as this is a convenient unit for infrared spectroscopy.

Table 6-2. Spectroscopic constants for a number of diatomic molecules are listed. The values are obtained from Reference [1] in the bibliography section at the end of this chapter. The term D_0 refers to the dissociation energy of the molecule.

	$\omega_e(cm^{-1})$	$B_e(cm^{-1})$	$\omega_e\chi_e(cm^{-1})$	$\alpha_0(cm^{-1})$	D_0 (eV)
$^{27}Al^1H$	1682.57	6.3962	29.145	0.188	<3.06
$^{27}Al^{35}Cl$	481.30	0.242	1.95	0.002	3.1
$^{11}B^1H$	2366	12.018	49	0.412	<3.51
$^{11}B^{14}N$	1514.6	1.666	12.3	0.025	5.0
$^{11}B^{16}O$	1885.44	1.7803	11.769	0.01648	9.1
$^{11}B^{19}F$	1390.8	1.518	11.3	0.017	4.3
$^{11}B^{35}Cl$	839.2	0.6838	5.11	0.00646	4.2
$^{11}B^{79}Br$	684.31	0.490	3.52	0.0035	4.1
$^{12}C^1H$	2861.6	14.457	64.3	0.534	3.47
$^{12}C^2H$	2101.0	7.808	34.7	0.212	3.52
$^{12}C^{14}N$	2068.705	1.8996	13.144	0.01735	4.5-7.6
$^{12}C^{16}O$	2170.21	1.9313	13.461	0.01748	***
$^{12}C^{32}S$	1285.1	0.8205	6.5	0.00624	7.8
$^{40}Ca^1H$	1299	4.2778	19.5	0.0963	$\leq$ 1.7
$^{40}Ca^{16}O$	650	0.464	6.6	0.006	5.9
$^1H^{19}F$	4138.52	20.939	90.069	0.770	$\leq$ 6.4
$^2H^{19}F$	2998.25	11.007	45.71	0.293	$\leq$ 6.4
$^1H^{35}Cl$	2989.74	10.5909	52.05	0.3019	4.430
$^2H^{35}Cl$	2090.78	5.445	52.05	0.3019	4.481
^1HBr	2649.67	8.473	45.21	0.226	3.75
$^1H^{127}I$	2309.5	6.551	39.73	0.183	3.056
$^{14}N^{16}O$	1904	1.7046	13.97	0.0178	5.29
$^{16}O^1H$	3735.21	18.871	82.81	0.714	4.35
$^{16}O^2H$	2720.1	10.01	44.2	0.29	4.39

6.7 TIME DEPENDENCE AND SELECTION RULES

Up to this point, the wavefunctions considered do not evolve with time. In some cases, the Hamiltonian may have time-dependent terms indicating that the system changes with time. An important example is when electromagnetic radiation interacts with a system. Electromagnetic radiation consists of electric and magnetic fields that oscillate in space and time. When electromagnetic radiation interacts with a molecule (such as in spectroscopy), the oscillating fields will result in a time-dependent element in the complete Hamiltonian for the molecule. As already observed in the case of infrared spectroscopy, this interaction may result in a transition of states.

If time, t, is a variable in a quantum mechanical system, then there must be an operator associated with time. The operator time, $\hat{t}$, (just like position) consists of multiplication by t.

$$\hat{t} = t \tag{6-62}$$

A Schroedinger equation is needed that describes a quantum mechanical system that includes the variable time (the ***time dependent Schroedinger Equation, TDSE***). As a result, there must be an operator such that when it operates on a wavefunction, it yields an energy eigenvalue. As can be confirmed via a dimensional analysis, the following operator $\hat{E}$ results in an energy eigenvalue when operated on an eigenfunction:

$$\hat{E} = i\hbar \frac{\partial}{\partial t}. \tag{6-63}$$

Postulate III (see Section 2.2) can now be generalized for any system including the variable time (Ψ is a wavefunction that includes time):

$$\hat{H}\Psi = \hat{E}\Psi \tag{6-64a}$$

or

$$\hat{H}\Psi = i\hbar \frac{\partial \Psi}{\partial t}. \tag{6-64b}$$

The wavefunction Ψ in Equation 6-64a and b is a function of time, and the Hamiltonian may also have time dependence. Based on the extension of Postulate III for the TDSE, the wavefunctions are eigenfunctions of both space and time.

The TDSE is a generalization of the time independent Schroedinger equation, TISE (the type of Schroedinger equation considered up to this point). The TDSE does not, however, invalidate the TISE. Rather, the TISE is a case where the Hamiltonian is independent of time. For a time independent Hamiltonian, the wavefunction Ψ is separable in terms of space and time.

$$\Psi(q,t) = \psi(q)\phi(t) \tag{6-65}$$

Equation 6-65 can now be substituted into the TDSE (Equation 6-64b).

$$\hat{H}\psi(q)\phi(t) = i\hbar\,\psi(q)\,\frac{\partial\phi(t)}{\partial t} \tag{6-66}$$

Dividing both sides of Equation 6-66 by $\psi(q)\phi(t)$ yields

$$\frac{\hat{H}\psi(q)}{\psi(q)} = i\hbar\,\frac{\{\partial\phi(t)\,/\,\partial t\}}{\phi(t)}\,. \tag{6-67}$$

Since both sides of Equation 6-67 depend on a different variable, the two sides of the equation must equal to the same energy eigenvalue, E. This results in the following:

$$\hat{H}\psi(q) = E\psi(q) \tag{6-68}$$

and

$$i\hbar\,\frac{\partial\phi(t)}{\partial t} = E\phi(t)\,. \tag{6-69}$$

Equation 6-68 is the TISE that has been used in this text so far. The solution to Equation 6-66 is

$$\phi(t) = e^{-\frac{iEt}{\hbar}}. \tag{6-70}$$

Since $\phi^*(t)\phi(t)$ is equal to one, then $\Psi^*\Psi = \psi^*(q)\psi(q)\phi^*(t)\phi(t) = \psi^*(q)\psi(q)$. In the case of a time independent Hamiltonian, the function $\phi(t)$ has no effect on the energy or particle distribution, hence it is ignored in time independent systems.

In systems where the Hamiltonian has time dependence (such as in the absorption of electromagnetic radiation), the separation of the TDSE into spatial and time differential equations is generally not possible or very difficult. One approach to solving these problems is to treat the time dependence in the Hamiltonian as a perturbation of the time independent Hamiltonian (***time-dependent perturbation theory***). The Hamiltonian for the system is separated into a time dependent part, H_t (the perturbation), and a time independent part, H_0. The TDSE is written as follows:

$$(\hat{H}_0 + \hat{H}_t)\Phi = i\hbar\frac{\partial\Phi}{\partial t}. \tag{6-71}$$

The wavefunction Φ is taken as a linear combination of the "unperturbed" stationary state wavefunctions Ψ_i (the product of the spatial wavefunction ψ_i and Equation 6-70.

$$\Phi = \sum_i a_i(t)\Psi_i \tag{6-72}$$

$$\Psi_i = \psi_i e^{-\frac{iE_it}{\hbar}} \tag{6-73}$$

The solution of Equation 6-71 involves solving for all of the $a_i(t)$ coefficients. Equations 6-72 and 6-73 can now be substituted into Equation 6-71.

$$\sum_i a_i(t)e^{-\frac{iE_it}{\hbar}}\hat{H}_0\psi_i + \sum_i \hat{H}_t a_i(t)e^{-\frac{iE_it}{\hbar}}\psi_i = i\hbar\sum_i \psi_i\frac{\partial}{\partial t}a_i(t)e^{-\frac{iE_it}{\hbar}} \tag{6-74}$$

Equation 6-74 can be simplified by recognizing that

$$\hat{H}_0\psi_i = E_i\psi_i$$

(the TISE) and that the derivative of the exponential on the right side of the equation is equal to a constant and the same exponential.

$$\sum_i a_i(t)e^{-\frac{iE_it}{\hbar}} E_i\psi_i + \sum_i \hat{H}_t a_i(t)e^{-\frac{iE_it}{\hbar}}\psi_i = \sum_i a_i(t)e^{-\frac{iE_it}{\hbar}} E_i\psi_i + i\hbar\sum_i \psi_i \frac{\partial a_i(t)}{\partial t}e^{-\frac{iE_it}{\hbar}}$$

$$\sum_i \hat{H}_t a_i(t)e^{-\frac{iE_it}{\hbar}}\psi_i = i\hbar\sum_i \psi_i \frac{\partial a_i(t)}{\partial t} e^{-\frac{iE_it}{\hbar}} \qquad (6\text{-}75)$$

Since the "unperturbed" Ψ_i wavefunctions are orthonormal, Equation 6-75 can be further simplified by multiplying both sides of the equation by Ψ_k^* and integrating over all spatial coordinates. The k can refer to any index.

$$\sum_i \left\langle \Psi_k \left| \hat{H}_t \right| a_i(t)\Psi_i \right\rangle = i\hbar\sum_i \frac{\partial a_i(t)}{\partial t} \left\langle \Psi_k \middle| \Psi_i \right\rangle \qquad (6\text{-}76)$$

The integral on the right of Equation 6-76 is non-zero (equal to one) only when k = i.

$$\sum_i \left\langle \Psi_k \left| \hat{H}_t \right| a_i(t)\Psi_i \right\rangle = i\hbar \frac{\partial a_k(t)}{\partial t} \qquad (6\text{-}77)$$

The integral in Equation 6-77 is solved for each k index to obtain each $a_k(t)$ coefficient.

In the case of spectroscopy, what is of interest is the moment at which a photon (the perturbation) is absorbed and whether it will result in a transition in states in the molecule (initial eigenstate i to a final eigenstate k). As a result, it is sufficient to concentrate on a short time interval defined as t = 0. In this case, the initial state Φ is only one stationary state.

$$a_{initial}(t = 0) = 1 \quad \text{and} \quad a_{i \neq initial}(t = 0) = 0$$

Equation 6-77 is then reduced to the following expression:

$$\left\langle \Psi_k \left| \hat{H}_t \right| \Psi_{initial} \right\rangle = i\hbar \frac{\partial a_k(t = 0)}{\partial t}. \tag{6-78}$$

The spectroscopic selection rules are determined by the integral on the left side of Equation 6-78. When the integral is zero, it is a ***forbidden transition*** between the initial and k stationary eigenstates. When the integral is non-zero, it is an ***allowed transition***.

$$\left\langle \Psi_k \left| \hat{H}_t \right| \Psi_{initial} \right\rangle = 0 \qquad \text{(forbidden transition)}$$

$$\left\langle \Psi_k \left| \hat{H}_t \right| \Psi_{initial} \right\rangle \neq 0 \qquad \text{(allowed transition)}$$

In the case of infrared spectroscopy, a transition from an initial v, J to a final v', J' vibrational and rotational eigenstates is either forbidden or allowed.

$$\left\langle \Psi_{v',J'} \left| \hat{H}_t \right| \Psi_{v,J} \right\rangle = 0 \qquad \text{(forbidden transition)} \tag{6-79a}$$

$$\left\langle \Psi_{v',J'} \left| \hat{H}_t \right| \Psi_{v,J} \right\rangle \neq 0 \qquad \text{(allowed transition)} \tag{6-79b}$$

The time dependent (perturbing) Hamiltonian consists of the interaction of the oscillating electric field of the impinging photon with the dipole moment of the molecule, μ. The electric field of a photon oscillates at a frequency ω_R.

$$\hat{H}_t = -\vec{\mu} \bullet [\vec{E}\cos(\omega_R t)]_{photon} \tag{6-80}$$

The dot product of the dipole moment of the molecule with the electric field of the photon arises because only certain orientations of the dipole moment of the molecule with the electric field of the photon result in a favorable interaction. Consider a diatomic molecule with a dipole in a constant electric field (i.e. between two charged plates, one positively charged and one negatively charged). When the positive end of the dipole is in line with the negative plate and the negative end of the dipole is in line with the positive plate (call this orientation 0°), the interaction is energetically favorable (negative energy). When the dipole of the molecule is oriented exactly opposite with the positive end in line with the positive plate and the negative end in line with the negative plate (call this orientation 180°), the interaction energy is equal in magnitude as before but now it is positive. When the positive and negative ends of the dipole of the molecule lie parallel to the positive and negative plates (orientation 90°), the interaction energy is zero. The orientation of the molecule with respect to its molecular axis varies with the rotational coordinate θ. Since the dipole moment of the molecule has a non-zero component only along the molecular axis, the dot product in Equation 6-80 varies with the cosine of θ. The dipole moment of the molecule is also related to the distance of separation between the atoms; hence, the interacting Hamiltonian in Equation 6-80 becomes:

$$\hat{H}_t = -\left|\vec{E}\right| \cos(\omega_R t)\mu(s) \cos\theta. \tag{6-81}$$

The selection rules in infrared spectroscopy arise from the integrals of the initial and final eigenstate wavefunctions with the interacting Hamiltonian in Equation 6-81.

The wavefunction for a particular vibrational/rotational eigenstate is a product of the radial and spherical harmonic wavefunctions. The v, J, and the degenerate M_J quantum numbers specify the state.

$$\psi_{vJM_J} = S_v(s)Y_{JM_J}(\theta,\phi)$$

Whether a specific transition is allowed or forbidden is determined by solving the integral in Equation 6-79.

$$\begin{aligned} &\left\langle \psi_{vJM_J} \middle| \hat{H}_t \middle| \psi_{v'J'M_J'} \right\rangle \\ &= -\left|\vec{E}\right| \cos(\omega_R t)\left\langle S_v(s)Y_{JM_J}(\theta,\phi) \middle| \mu(s)\cos\theta \middle| S_{v'}(s)Y_{J'M_J'}(\theta,\phi) \right\rangle \end{aligned} \tag{6-82}$$

The integral in Equation 6-82 can be factored into two integrals, one over the radial coordinate and the other over the angular coordinates.

$$\left\langle \psi_{vJM_J} \left| \hat{H}_t \right| \psi_{v'J'M_J'} \right\rangle$$
$$= -\left|\vec{E}\right| \cos(\omega_R t) \left\langle S_v(s) \left| \mu(s) \right| S_{v'}(s) \right\rangle \left\langle Y_{JM_J}(\theta, \phi) \left| \cos\theta \right| Y_{J'M_J'}(\theta, \phi) \right\rangle \quad (6\text{-}83)$$

The integral over the angular coordinates can be evaluated for either a specific transition from an initial J, M_J eigenstate to a final J', M_J' eigenstate explicitly or, as in the equation below, for any transition by using the recursion relationship for the Legendre polynomials.

$$\left\langle Y_{JM_J}(\theta, \phi) \left| \cos\theta \right| Y_{J'M_J'}(\theta, \phi) \right\rangle$$
$$= \delta_{M_J M_J'} \sqrt{\delta_{J,J'+1} \frac{J^2 - M_J^2}{4J^2 - 1} + \delta_{J,J'-1} \frac{J'^2 - M_J^2}{4J'^2 - 1}} \quad (6\text{-}84)$$

The δ's in Equation 6-84 refer to the orthonormality integrals for the wavefunctions as follows:

$$\delta_{M_J M_J'} = \left\langle \psi_{M_J}(\phi) \left| \psi_{M_J'}(\phi) \right. \right\rangle = 0 \quad (\text{If } M_J \neq M_J')$$
$$= 1 \quad (\text{If } M_J = M_J');$$

$$\delta_{J,J'+1} = \left\langle Y_{JM_J}(\theta, \phi) \left| Y_{J'+1,M_J}(\theta, \phi) \right. \right\rangle = 0 \quad (\text{If } J \neq J' + 1)$$
$$= 1 \quad (\text{If } J = J' + 1);$$

and

$$\delta_{J,J'+1} = \left\langle Y_{JM_J}(\theta, \phi) \left| Y_{J'-1,M_J}(\theta, \phi) \right. \right\rangle = 0 \quad (\text{If } J \neq J' - 1)$$
$$= 1 \quad (\text{If } J = J' - 1).$$

As a result, the integral in Equation 6-83 is non-zero (allowed transition) if the quantum state M_J *does not change and the J quantum state increases or decreases by one.*

In order to solve the integral over the radial coordinate, a function is needed to describe how the dipole of the molecule varies with the displacement of the distance separation of the atoms, s. One method is to represent the dipole as a power series expansion from the equilibrium (at s = 0) dipole moment of the molecule, μ_e.

$$\mu(s) = \mu_e + s\frac{d\mu}{ds}\bigg|_{s=0} + \tfrac{1}{2}s^2\frac{d^2\mu}{ds^2}\bigg|_{s=0} + \cdots \tag{6-85}$$

Equation 6-85 is now substituted into the radial integral. This integral can be solved for any $v \to v'$ transition if the vibrational motion is assumed to be harmonic and by utilizing the Hermite polynomial recursion relationship.

$$\begin{aligned}\langle S(s)|\mu(s)|S(s)\rangle &= \mu_e\delta_{vv'} \\ &+ \frac{d\mu}{ds}\bigg|_{s=0}\sqrt{\frac{\hbar}{2\sqrt{mk}}}(\sqrt{v}\delta_{v,v'+1} + \sqrt{v'}\delta_{v+1,v'}) \\ &+ \tfrac{1}{2}\frac{d^2\mu}{ds^2}\bigg|_{s=0}\frac{\hbar}{2\sqrt{mk}}[(2v+1)\delta_{v,v'} \\ &+ \sqrt{v(v-1)}\delta_{v,v'+2} + \sqrt{v'(v'-1)}\delta_{v+2,v'}] + \cdots\end{aligned} \tag{6-86}$$

Equation 6-86 must be analyzed term by term. The δ's refer to the orthonormality integrals of the harmonic oscillator wavefunctions. The first term is non-zero when the vibrational quantum state does not change but only when the molecule has a non-zero equilibrium dipole moment (such as HCl and HF). The second term is non-zero either when the vibrational quantum state increases or decreases by one if the first derivative of the dipole moment of the molecule is non-zero. The third term is non-zero when the vibrational quantum state increases or decreases by two if the second derivative of the dipole moment of the molecule is non-zero. Further expansion terms reveal that any change in vibrational states is allowed for a vibrational transition for a molecule with a non-zero equilibrium dipole moment. The vibrational transition observed in an infrared spectrum must be determined using the Boltzmann distribution that was described previously.

In summary, an allowed transition occurs with a diatomic molecule that has a non-zero dipole moment and the rotational state changes by one ($\Delta J = \pm 1$) and the degenerate M_J state does not change. The vibrational state can change by any value; however, the frequencies in the spectrometer scan dictate the actual vibrational transition.

REFERENCES

[1] G. Herzberg, *Molecular Spectra and Molecular Structure. Infrared Spectra of Diatomic Molecules*, Van Nonstrand Reinhold, New York, 1950.

[2] C. Dykstra, *Quantum Chemistry and Molecular Spectroscopy*, Prentice Hall, Englewood Cliffs, New Jersey, 1992.

[3] W.H. Flygare, *Molecular Structure and Dynamics*, Prentice Hall, Englewood Cliffs, New Jersey, 1978.

PROBLEMS AND EXERCISES

6.1) A molecule absorbs a photon with a wavelength of 120.3 nm. Determine the following: (a) the energy difference between the initial and final quantum states of the molecule involved in the transition; and (b) the energy required to cause 1.00 mole of molecules to undergo this transition.

6.2) Explicitly verify Equation 6-40a and 6-40b using second-order perturbation theory.

6.3) Estimate the ground-state vibrational energy of $^{1}H^{35}Cl$ using Equation 6-41 and the data in Table 6-2. How does this compare to the ground vibrational energy of $^{2}H^{35}Cl$ obtained in the same way? What can you conclude about the effect of isotopic substitution and bonding?

6.4) From the data in Table 6-2, determine the equilibrium bond length for $^{12}C^{16}O$. Estimate the equilibrium rotational constant for $^{12}C^{18}O$.

6.5) Sketch an infrared spectrum of $^{1}H^{19}F$ at 300K and at 1000K using the harmonic oscillator and rigid rotor approximations.

6.6) Correct the sketch above taking into account vibrational anharmonicity, centrifugal distortion, and vibration-rotation coupling.

6.7) Make a plot of the vibrational energy levels of $^{1}H^{19}F$ in terms of the harmonic oscillator and the Morse Potential as in Figure 6-6.

Chapter 7

Vibrational and Rotational Spectroscopy of Polyatomic Molecules

In the previous chapter, vibrational/rotational (i.e. infrared) spectroscopy of diatomic molecules was analyzed. The same analysis is now applied to polyatomic molecules. Polyatomic molecules have more than one bond resulting in additional vibrational degrees of freedom. Rotation of linear polyatomic molecules is mechanically equivalent to that of diatomic molecules; however, the rotation of non-linear polyatomic molecules results in more than one degree of rotational freedom. The result of the additional vibrational and rotational degrees of freedom for polyatomic molecules is to complicate the vibrational/rotational spectra of polyatomic molecules relative to spectra of diatomic molecules. Though the spectra of polyatomic molecules are more complicated, many of the same features exist as in the spectra of diatomic molecules. As a result, a similar approach will be used in this chapter. The mechanics of a model system will be solved, determine the selection rules, and the features of a spectrum will be predicted.

7.1 ROTATIONAL SPECTROSCOPY OF LINEAR POLYATOMIC MOLECULES

The rotation of a linear polyatomic molecule can be viewed as an infinitesimally thin rod rotating about its one axis of rotation. This is mechanically equivalent to the rotation of a diatomic molecule. The only difference is how the moment of inertia is determined now that there are more than two masses.

$$I = \frac{\sum_{i<j} m_i m_j r_{ij}^2}{\sum_i m_i} \tag{7-1}$$

The term r_{ij} refers to the distance between the i and j atoms in the polyatomic molecule. The summation in the numerator is between *each pair of atoms* in the molecule. The summation in the denominator is the sum over all of the atoms in the molecule.

Example 7-1

Problem: Calculate the moment of inertia for $^1H^{12}C^{14}N$ given that the normal bond lengths of the H-C and the C≡N bonds are 0.1066 and 0.1156 nm respectively.

Solution: Equation 7-1 is expanded for $^1H^{12}C^{14}N$. The values are then substituted into the expanded expression. The mass of each atom is determined in kg.

$$m_{^1H} = 1amu * 1.6604x10^{-27}\,kg/amu = 1.6604x10^{-27}\,kg$$

$$m_{^{12}C} = 12amu * 1.6604x10^{-27}\,kg/amu = 1.6604x10^{-27}\,kg$$

$$m_{^{14}N} = 14amu * 1.6604x10^{-27}\,kg/amu = 1.6604x10^{-27}\,kg$$

Since the atoms are viewed as point masses with the mass concentrated in the nuclei, the distance r_{HN} is equal to the sum of the H-C and C≡N normal bond lengths, $r_{HN} \cong r_{HC} + r_{CN}$.

$$I = \frac{m_H m_C r_{HC}^2 + m_H m_N r_{HN}^2 + m_C m_N r_{CN}^2}{m_H + m_C + m_N} = 1.8937x10^{-46}\,kg \cdot m^2$$

Due to the mechanical equivalence, the Schroedinger equation and the resulting rotational energy eigenvalues is the same for linear polyatomic molecules as for diatomic molecules.

$$-\frac{\hbar^2}{2I}\hat{L}^2 Y_{Jm_J}(\theta,\phi) = E_J Y_{Jm_J}(\theta,\phi)$$

$$E_J = J(J+1)B\,; \qquad B = \frac{\hbar^2}{2I} \tag{7-2}$$

The only difference in the solution for a linear polyatomic molecule compared to that for a diatomic molecule is that the moment of inertia I is given by Equation 7-1. The eigenfunctions Y_{Jm_J} are the spherical harmonics.

The selection rules for rotational transitions in linear polyatomic molecules are also the same as for diatomic molecules. The transition ΔJ is equal to ±1 in infrared spectroscopy and +1 in purely rotational spectroscopy (i.e. microwave spectroscopy) but only if the molecule has a non-zero dipole moment (see Section 6.7). A rotational transition for H-C≡C-Cl will be observed whereas for H-C≡C-H no rotational transition will be observed due to its zero dipole moment.

The rotational transitions for linear polyatomics in microwave spectroscopy will be J = 0 → J = 1, J = 1 → J = 2, and so on. The corresponding energy differences for a rotational transition of J = 0 → J = 1

Table 7-1. The rotational constants and natural abundance of several isotopic forms of the linear molecule OCS obtained from the spectrum in Figure 7-1 are presented in this table. Since each absorption line, $\Delta E(MHz)$, in the spectrum corresponds to a J = 0 → J = 1 transition, $\Delta E(MHz) = 2B$. Data obtained from W.H. Flygare, *Molecular Structure and Dynamics*, Prentice-Hall, New Jersey, 1978.

Isotopic Species	B (MHz)	Natural Abundance, %
$^{16}O^{12}C^{32}S$	6,081.49	94.00
$^{16}O^{13}C^{32}S$	6,061.89	1.00
$^{17}O^{12}C^{32}S$	5,883.67	0.04
$^{18}O^{12}C^{32}S$	5,704.83	0.20
$^{16}O^{12}C^{34}S$	5,932.82	4.00
$^{16}O^{12}C^{33}S$	6,004.91	0.72
$^{16}O^{13}C^{34}S$	5,911.73	0.04

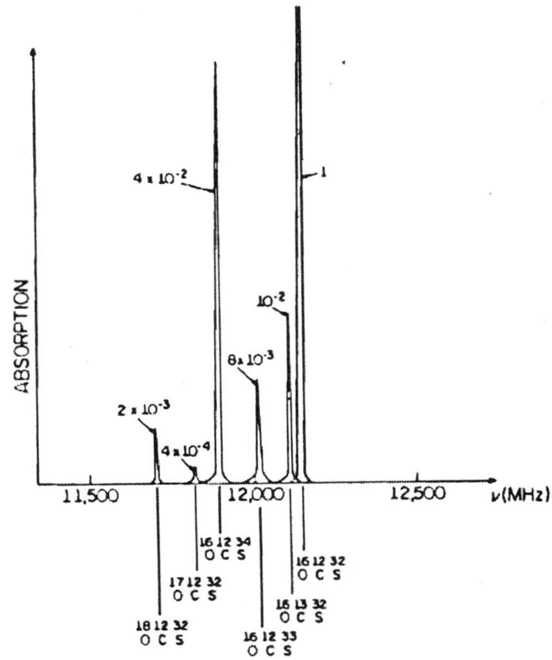

Figure 7-1. A microwave spectrum showing the relative intensities of some of the naturally occurring isotopes of the linear molecule OCS as observed by the relative intensities of the J = 0 → J = 1 transitions. The rotational constants derived for each isotopic form is listed in Table 7-1. Reproduced with permission from W.H. Flygare, *Molecular Structure and Dynamics*, Prentice-Hall, New Jersey, 1978.

is 2B; for a rotational transition of J = 1 → J = 2 it is 4B; and so on just as in diatomic molecules. The energy difference between rotational levels (2B, 4B, and so on) is equal to the energy of the absorbed microwave photon. The value of B is inversely proportional to the moment of inertia, and the moment of inertia is directly proportional to bond lengths. As a result, the energy of a rotational transition can be used to obtain the lengths of the bonds in the polyatomic molecule as in the case of diatomic molecules. However, because there is more than one bond in a polyatomic molecule, one value for the rotational constant B is not sufficient to determine the multiple bond lengths in the molecule.

A linear polyatomic molecule with N + 1 atoms has a total of N bond lengths to be determined. In order to determine N bond lengths (in other words N unknowns), a total of N equations are needed. The N equations can be obtained by measuring the rotational constants for N isotopically substituted forms of the molecule. The isotopically substituted molecules will each have different values for the rotational constant B as a result of the different masses of the atoms. The bond distances are assumed to be the same regardless of the isotopic substitution since the number of neutrons in the nuclei should not affect the chemical bonding. This assumption is only valid for equilibrium structures of the molecule. The increased mass does have a slight affect on the zero-point average vibrational bond length that can be corrected for by a more detailed analysis. Because of this assumption, the structure obtained for the molecule via isotopic substitution is called the ***substitution structure*** to distinguish it from other methods of structural determination such as crystallography.

The structure can be elucidated from the rotational constants of the various naturally occurring isotopic forms of the molecule. In the case of a linear molecule such as OCS for which the data is presented in Table 7-1, the substitution structure can be determined by two different isotopic structures (since there are two unknown bond lengths to be determined, C=O and C=S).

Example 7-2

Problem: Using the data from Table 7-1, determine the substitution structure of OCS.

Solution: Since there are only two bonds, only two equations are needed to obtain the bond distances r_{CO} and r_{CS}. The two equations are taken from the two most naturally abundant isotopic forms, $^{16}O^{12}C^{32}S$ and $^{16}O^{12}C^{34}S$. The moment of inertia for each isotopic form is determined from its respective rotational constant.

$$I(^{16}O^{12}C^{32}S) = \frac{\hbar^2}{2hB_{^{16}O^{12}C^{32}S}} = 1.37990x10^{-45}\,kg \cdot m^2$$

$$I(^{16}O^{12}C^{34}S) = 1.41450x10^{-45}\,kg\cdot m^2$$

Using Equation 7-1, the moment inertia for each isotopic form of OCS becomes:

$$I(^{16}O^{12}C^{32}S) = \frac{m_{^{12}C}m_{^{16}O}r_{CO}^2 + m_{^{16}O}m_{^{32}S}r_{OS}^2 + m_{^{12}C}m_{^{32}S}r_{CS}^2}{m_{^{16}O} + m_{^{12}C} + m_{^{32}S}}$$

$$I(^{16}O^{12}C^{34}S) = \frac{m_{^{12}C}m_{^{16}O}r_{CO}^2 + m_{^{16}O}m_{^{34}S}r_{OS}^2 + m_{^{12}C}m_{^{34}S}r_{CS}^2}{m_{^{16}O} + m_{^{12}C} + m_{^{34}S}}$$

Since the atoms are viewed as point-masses, $r_{OS} \cong r_{CO} + r_{CS}$ (see Example 7-1). This results in the following two equations once the masses of the atoms are substituted into the moment of inertia expressions.

$$I(^{16}O^{12}C^{32}S) = 1.37990x10^{-45}\,kg\cdot m^2 = 1.9484x10^{-26}\,kg\cdot r_{CO}^2 + 2.4797x10^{-26}\,kg\cdot r_{CS}^2 + 1.4170x10^{-26}\,kg\cdot r_{CS}\cdot r_{CO}$$

$$I(^{16}O^{12}C^{34}S) = 1.41450x10^{-45}\,kg\cdot m^2 = 1.9713x10^{-26}\,kg\cdot r_{CO}^2 + 2.5498x10^{-26}\,kg\cdot r_{CS}^2 + 1.4570x10^{-26}\,kg\cdot r_{CS}\cdot r_{CO}$$

The two equations above are solved simultaneously resulting in the bond distances r_{CO} and r_{CS} using the moments of inertia from the isotopic forms $^{16}O^{12}C^{32}S$ and $^{16}O^{12}C^{34}S$. A more thorough analysis would involve averaging over the bond distances obtained from all of the isotopic forms given in Table 7-1.

$$r_{CO} = 1.34992x10^{-10}\,m = 0.134992nm$$

$$r_{CS} = 1.68353x10^{-10}\,m = 0.168353nm$$

The bond lengths of any N + 1 size linear polyatomic molecule can be obtained via microwave spectroscopy given the spectra of N different isotopic forms of the molecule. The computational complexity increases though with increasing N.

7.2 ROTATIONAL SPECTROSCOPY OF NON-LINEAR POLYATOMIC MOLECULES

Since a molecule rotates about its center of mass, it is convenient to define the coordinates for the axes of rotation of a molecule about its center of mass. The first step is to determine the center of mass of the molecule. The point that the center of mass of a molecule is located is defined as X, Y, Z. The center of mass of a molecule is determined as follows:

$$\sum_{i}^{N+1} m_i(x_i - X) = 0; \quad \sum_{i}^{N+1} m_i(y_i - Y) = 0; \quad \sum_{i}^{N+1} m_i(z_i - Z) = 0 \quad (7\text{-}3)$$

where N + 1 corresponds to the total number of nuclei in the molecule. The terms x_i, y_i, and z_i correspond to the positions of each nuclei in an x,y,z coordinate system.

Example 7-3

Problem: Determine the center of mass (X, Y, Z) of water. The bond angle is 104.5° and the bond distance of O-H is 0.0958 nm.

Solution: The structure of a water molecule is shown in the figure below.

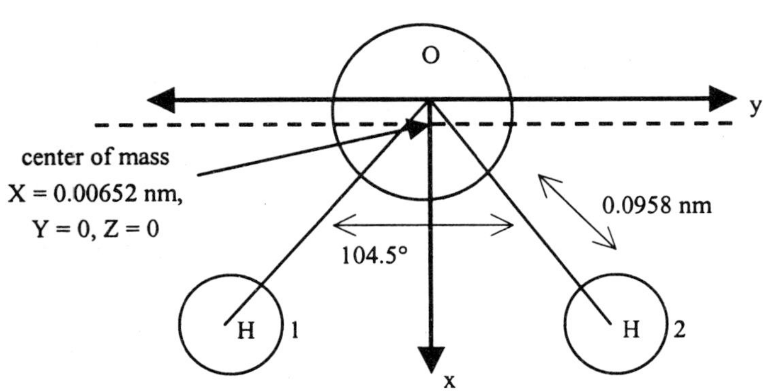

The values for X,Y, and Z are solved for by using each relationship in Equation 7-3.

$$\sum_{i}^{3} m_i(x_i - X) = m_H(x_{H(1)} - X) + m_H(x_{H(2)} - X) + m_O(x_o - X) = 0$$

$$m_H x_{H(1)} + m_H x_{H(2)} + m_O x_O = (2m_H + m_O)X$$

$$X = \frac{m_H(x_{H(1)} + x_{H(2)}) + m_O x_O}{(2m_H + m_O)}$$

Based on the coordinate system used in the diagram, the oxygen atom is at the origin; hence $x_O = 0$. The position of each hydrogen atom is given below.

$$x_{H(1)} = x_{H(2)} = 0.0958nm\left(\cos\left(\frac{104.5^\circ}{2}\right)\right) = 0.0587nm$$

Substituting the atomic masses into the expression for X results in the position for the center of mass along the x-axis.

$$X = \frac{2(1.6604x10^{-27}\,kg)(0.0587nm)}{2.9887x10^{-26}\,kg} = 6.52x10^{-3}\,nm$$

The center of mass along the y-axis is found in a similar manner. The center of mass along the z-axis is zero since the molecule is on the x-y plane.

$$Y = \frac{m_H[\sin(-52.25^\circ) + \sin(52.25^\circ)]}{(2m_H + m_O)} = 0$$

Note that the center of mass of a water molecule is very close to the oxygen atom due to its large mass relative to that of the hydrogen atoms. The center of mass is shown in the figure for this example.

The coordinate system is now drawn such that the origin is at the center of mass of the molecule. In order to describe the rotational motion of a non-linear molecule, three angular coordinates are needed resulting in three moments of inertia. The position of each atom is now expressed in a coordinate system whereby the center of mass of the molecule is at the origin and each atom is along the axes labeled by convention as a, b, and c. This coordinate system is called the ***principal inertial axis system***. The three moments of inertia that result from the principal inertial axis system are called the ***principal moments of inertia***.

$$I_a = \sum_i^{N+1} m_i (b_i^2 + c_i^2) \tag{7-4}$$

$$I_b = \sum_i^{N+1} m_i (a_i^2 + c_i^2) \tag{7-5}$$

$$I_c = \sum_i^{N+1} m_i (a_i^2 + b_i^2) \tag{7-6}$$

By convention, the axes are named such that $I_a \leq I_b \leq I_c$.

Example 7-4

Problem: Determine the three principal moments of inertia for a water molecule.

Solution: The center of mass of the molecule was previously determined in Example 7-3. The location of each atom relative to the center of mass is determined as follows (see the figure in Example 7-3).

H(1): $$\left(-(0.0958nm)\sin(52.25^\circ), (0.0958)\cos(52.25^\circ) - 0.00652, 0\right) = (-0.0757nm, 0.0521nm, 0)$$

H(2): $(0.0757nm, 0.0521nm, 0)$

O: $(0, -0.00652nm, 0)$

The principal moments of inertia follow directly from Equations 7-4 through 7-6.

$$I_a = 2m_H(0.0521nm)^2 + m_O(-0.00652nm)^2 = 0.00612amu \cdot nm^2$$
$$= 1.02x10^{-47}\,kg \cdot m^2$$

$$I_b = 2m_H(0.0757nm)^2 = 0.0115amu \cdot nm^2 = 1.91x10^{-47}\,kg \cdot m^2$$

$$I_c = 2m_H[(0.0521nm)^2 + (0.0757nm)^2] + m_O(-0.00652nm)^2$$
$$= 0.0176amu \cdot nm^2 = 2.92x10^{-47}\,kg \cdot m^2$$

Symmetry in this problem leads to the correct designation of the a, b, and c axes. For non-symmetrical molecules, it is more difficult to determine the unique orientation that results in the correct designation for I_a, I_b, and I_c.

For a linear molecule, one of the rotational axes will lie along the atomic axis. By convention this is designated as the a-axis since this will make the moment of inertia along the a-axis, I_a, equal to zero. The values for b_i and c_i will equal to zero for all the atoms since they run along the a-axis only. By inspection of Equations 7-4 through 7-6, the following principal moments of inertia for a linear molecule will result.

$$\text{linear molecules:} \qquad I_a = 0; \qquad I_b = I_c \qquad (7\text{-}7)$$

Equation 7-7 confirms the results from Section 7-1 (and from Chapter 6 for diatomic molecules) that linear molecules have doubly degenerate rotational states designated as J and M_J. There is a single rotational constant for linear molecules that is defined in the same way as in Equation 7-2.

If a non-linear molecule has principal moments of inertia that are equal and the third is non-zero, then it is a ***symmetric top molecule***. There are two types of symmetric top molecules: ***prolate symmetric top*** and ***oblate symmetric top*** molecules.

$$\text{prolate symmetric top:} \qquad I_c = I_b > I_a \qquad (7\text{-}8)$$

oblate symmetric top: $I_a = I_b > I_c$ (7-9)

A prolate symmetric top molecule has most of its mass spread along a high symmetry axis. A prolate symmetric top can be envisioned to be similar to a baseball bat. Examples of prolate symmetric tops include CH_3F and $CH_3C{\equiv}CH$. Additional examples are given in Table 7-2. An oblate symmetric top molecule has most of its mass spread out over a plane. An oblate symmetric top can be envisioned as a discus. Examples of oblate symmetric top molecules include benzene and CF_3H. Additional examples are given in Table 7-3.

Another possibility is that a molecule has all three principal moments of inertia being equal. In this case, the molecule is called a ***spherical top***.

spherical top: $I_a = I_b = I_c$ (7-10)

Examples of spherical top molecules include tetrahedral molecules such as CH_4 and octahedral molecules such as SF_6.

The most common case is when all of the three principal moments of inertia are not equal. This type of molecule is called an ***asymmetric top***.

asymmetric top: $I_a < I_b < I_c$ (7-11)

As can be seen by Example 7-4, a water molecule is an example of an asymmetric top molecule.

Table 7-2. The principal moments of inertia in 10^{-47} kg·m^2 along with the rotational constants in MHz for a number of prolate symmetric top molecules ($I_a < I_b = I_c$) are listed in the table below. Data obtained from W.H. Flygare, *Molecular Structure and Dynamics*, Prentice-Hall, New Jersey, 1978.

	I_a	$I_b = I_c$	A	B
FCH_3	5.300	32.864	158,319.0	25,536.12
$ClCH_3$	5.310	63.132	158,020.9	13,292.95
$BrCH_3$	5.320	87.708	157,723.8	9,658.19
ICH_3	5.351	111.876	156,839.0	7,501.31
$CH_3C{\equiv}CCH_3$	5.32	98.202	157,723.8	8,545.84

Table 7-3. The principal moments of inertia in 10^{-47} kg·m^2 along with the rotational constants in MHz for some oblate symmetric top molecules ($I_a = I_b < I_c$) are listed in the table below. Data obtained from W.H. Flygare, *Molecular Structure and Dynamics*, Prentice-Hall, New Jersey, 1978.

	$I_a = I_b$	I_c	A	C
HCF_3	81.093	149.122	10,384.74	5,627.70
NF_3	78.571	87.333	10,680.96	9,609.37
$BrCH_3$	107.318	111.627	7,819.90	7,518.06

A non-linear molecule will have three orthogonal rotations along the a, b, and c rotational axes. This results in three rotational constants being defined as follows.

$$A = \frac{\hbar^2}{2I_a} \qquad (7\text{-}12)$$

$$B = \frac{\hbar^2}{2I_b} \qquad (7\text{-}13)$$

$$C = \frac{\hbar^2}{2I_c} \qquad (7\text{-}14)$$

Based on the convention that $I_a \leq I_b \leq I_c$, the rotational constants will have the following ordering: $A \geq B \geq C$. The rotational constants can be ordered into a single, dimensionless number called ***Ray's asymmetry parameter***, κ. The Ray's asymmetry parameter scales asymmetric top molecules between prolate and oblate limits.

$$\kappa = \frac{2B - A - C}{A - C} \qquad (7\text{-}15)$$

For a prolate symmetric top molecule, $B = C$ and $\kappa = -1$. For an oblate symmetric top molecule, $A = B$ and $\kappa = +1$. All asymmetric top molecules will fall in between, $-1 < \kappa < +1$. For asymmetric molecules where $\kappa \cong -1$,

Table 7-4. The rotational constants in MHz along with the Ray's asymmetry parameter, κ, for some near prolate molecules (A > B ≅ C) are shown. Data obtained from W.H. Flygare, *Molecular Structure and Dynamics*, Prentice-Hall, New Jersey, 1978.

	A	B	C	κ
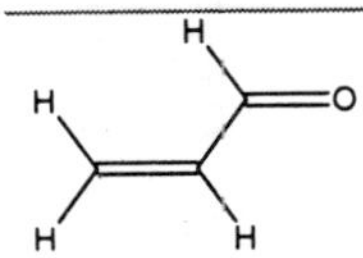	47,353.3	4,659.44	4,242.79	-0.98
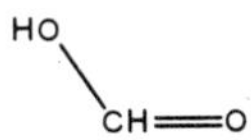	77,510.7	12,055.0	10,416.2	-0.95
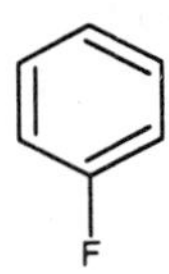	5,663.5	2,570.64	1,767.94	-0.59

the molecule is termed ***near prolate***. For asymmetric molecules where κ ≅1, the molecule is termed ***near oblate***. Some examples of near prolate molecules are listed in Table 7-4. The energy eigenvalues for near prolate and near oblate can be obtained from the solution to the Schroedinger equation for prolate and oblate molecules by treating the difference as a perturbation.

The Hamiltonian for the rotational motion of a non-linear polyatomic molecule can now be written. The rotational motion is free so the Hamiltonian will have only kinetic energy operators along the a, b, and c rotational axes.

$$\hat{H} = \frac{\hat{J}_a^2}{2I_a} + \frac{\hat{J}_b^2}{2I_b} + \frac{\hat{J}_c^2}{2I_c} \tag{7-16}$$

Equation 7-16 can now be rearranged to produce a term with the total angular momentum.

$$\hat{H} = \frac{\hat{J}_a^2 + \hat{J}_b^2 + \hat{J}_c^2}{2I_b} + \hat{J}_a^2\left(\frac{1}{2I_a} - \frac{1}{2I_b}\right) + \hat{J}_c^2\left(\frac{1}{2I_c} - \frac{1}{2I_b}\right)$$

$$\hat{H} = \frac{\hat{J}^2}{2I_b} + \hat{J}_a^2\left(\frac{1}{2I_b} - \frac{1}{2I_c}\right) + \hat{J}_c^2\left(\frac{1}{2I_c} - \frac{1}{2I_b}\right) \qquad (7\text{-}17)$$

If a molecule is a spherical top ($I_a = I_b = I_c$), only the total angular momentum term in Equation 7-17 will remain.

$$\hat{H} = \frac{\hat{J}^2}{2I_b} \qquad (7\text{-}18)$$

The Hamiltonian for spherical top molecules is the same form as for linear polyatomic molecules and diatomic molecules. The energy eigenvalues are the same form, and the microwave spectra will also be similar.

For prolate ($I_c = I_b > I_a$) and oblate ($I_a = I_b > I_c$) molecules, two of the moments of inertia are the same canceling one term from the Hamiltonian in Equation 7-17. This results in two different Hamiltonians, one for prolate molecules and the other for oblate molecules.

$$\hat{H}^{prolate} = \frac{\hat{J}^2}{2I_b} + \hat{J}_a^2\left(\frac{1}{2I_a} - \frac{1}{2I_b}\right) \qquad (7\text{-}19)$$

$$\hat{H}^{oblate} = \frac{\hat{J}^2}{2I_b} + \hat{J}_c^2\left(\frac{1}{2I_c} - \frac{1}{2I_b}\right) \qquad (7\text{-}20)$$

The form of the Hamiltonians in Equations 7-19 and 7-20 effectively states that taking the square of the total angular momentum operator, $\hat{J}^2$, and a squared component momentum operator, $\hat{J}_a^2$ or $\hat{J}_c^2$, applied to the eigenfunction yields the energy eigenvalues. The result obtained will be similar to the result obtained for the Particle-on-a-Sphere problem in Equations 3-26 and 3-28. When the total squared angular momentum operator is applied to the wavefunction, the result is proportional to $J(J + 1)$. When the conjugate component angular momentum operator is applied to the wavefunction, the result is proportional to a constant times another quantum number - namely M_J for the $\hat{L}_z$ operator (or M_J^2 since the conjugate component operator is squared in this case). The solution for the energy eigenvalues can be drawn from this similar set of eigenequations from the Particle-on-a-Sphere problem. The energy eigenvalues for the free rotation of prolate and oblate molecules will depend on two quantum numbers, J and K.

$$E_{J,K}^{prolate} = J(J+1)\frac{\hbar^2}{2I_b} + K^2\left(\frac{\hbar^2}{2I_a} - \frac{\hbar^2}{2I_b}\right) \quad (7\text{-}21)$$

$$E_{J,K}^{oblate} = J(J+1)\frac{\hbar^2}{2I_b} + K^2\left(\frac{\hbar^2}{2I_c} - \frac{\hbar^2}{2I_b}\right) \quad (7\text{-}22)$$

Equations 7-21 and 7-22 can be rewritten in terms of the rotational constants.

$$E_{J,K}^{prolate} = J(J+1)B + K^2(A-B) \quad (7\text{-}23)$$

$$E_{J,K}^{oblate} = J(J+1)B + K^2(C-B) \quad (7\text{-}24)$$

The eigenvalues needed to completely specify the rotational state of an oblate or prolate molecule are as follows (M_J is degenerate).

$$J = 0,1,2,3,...$$

$$M_J = -J, -J+1, -J+2, ..., 0, 1, 2, ..., J$$

$$K = -J, -J+1, -J+2, ..., 0, 1, 2, ..., J$$

An energy level diagram for an oblate top molecule is shown in Figure 7-2.

The selection rules for the rotational spectra of symmetric top molecules are as follows. Note that the degenerate rotational quantum number M_J can now change as a result of a transition.

$$\Delta J \pm 1\ ; \qquad \Delta K = 0\ ; \qquad \Delta M_J = \pm 1$$

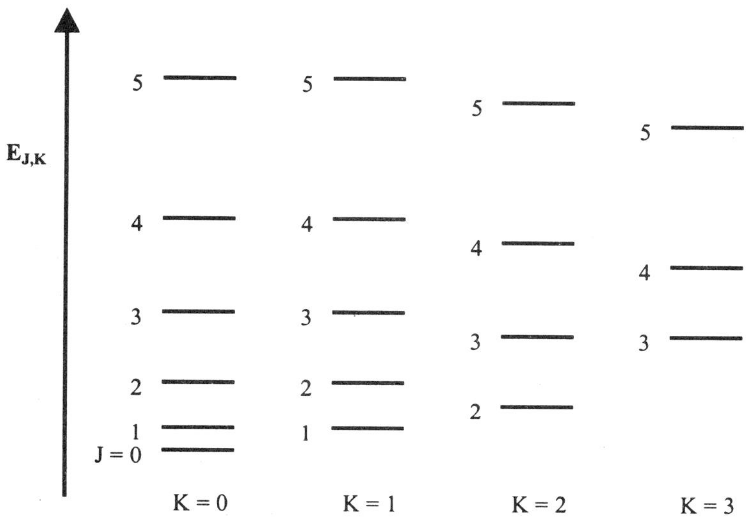

Figure 7-2. A rotational energy level diagram for an oblate symmetric top molecule is shown. Each column represents a K manifold of rotational J energy levels. In a rotational spectrum, allowed transitions occur within a particular K manifold because of the selection rule $\Delta K = 0$.

Rotational transitions in a spectrum will follow along a particular K manifold of rotational energy levels; however, at a given temperature a system of molecules may occupy more than one K level. As can be seen in Figure 7-2, since the various rotational states in a particular K manifold are close to rotational states in other K manifolds, the rotational spectrum can be expected to be very complicated with possible absorptions overlapping even in highly resolved spectra.

Another important mode of rotation in polyatomic molecules is ***internal modes of rotation***. As an example, consider the rotation of a methyl group about the C-C bond axis in ethane. The rotation of the methyl group can be approximated as a free rotor about the ϕ angle as in the Particle-on-a-Ring model problem (see Section 3.1). From the moment of inertia of the methyl group, the energy of the internal rotational states can be obtained from Equation 3-6.

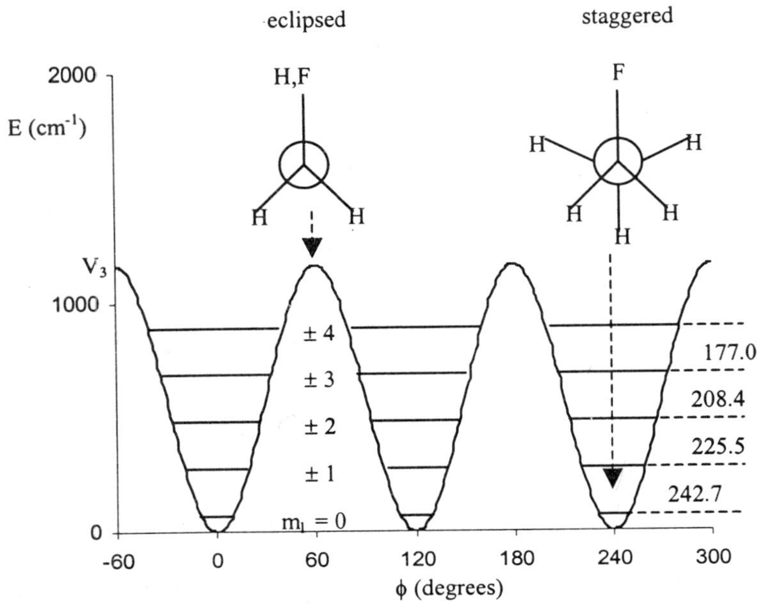

Figure 7-3. The barriers to internal rotation for ethyl fluoride are shown. The value for V_3 is 1158 cm^{-1}. Data obtained from G. Sage and W. Klemperer, *J. Chem. Phys.*, **39**, 371 (1963).

$$E = m_l^2 \frac{\hbar^2}{2I} = m_l^2 (10.6 cm^{-1})$$

If the methyl groups in ethane behave as free rotors, the transitions between eigenstates would be in multiples of 10.6 cm^{-1}.

The free rotor model is not adequate for most molecules as there is in general some potential barrier to internal rotation. As an example, consider ethyl fluoride. Ethyl fluoride, just like ethane, has a predominately three-fold potential barrier as shown in Figure 7-3. The potential barrier can be expressed as the following function.

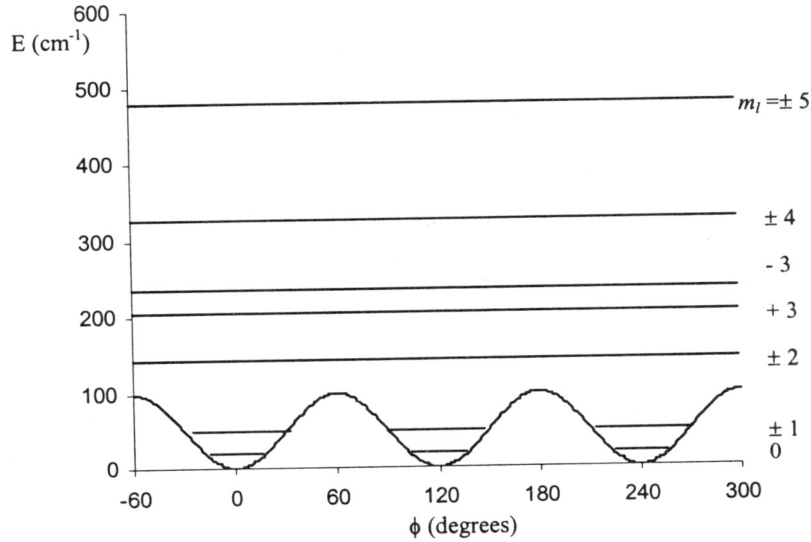

Figure 7-4. Internal rotation energy levels for a molecule with a barrier hindering rotation of 100 cm^{-1} and $E^{(0)} = m_l^2$ (17 cm^{-1}). Note that if the molecule is in the $m_l = 0$ or ±1, the molecule is classically locked into a particular configuration. Also note that the $m_l = \pm 3$ eigenstates are no longer degenerate.

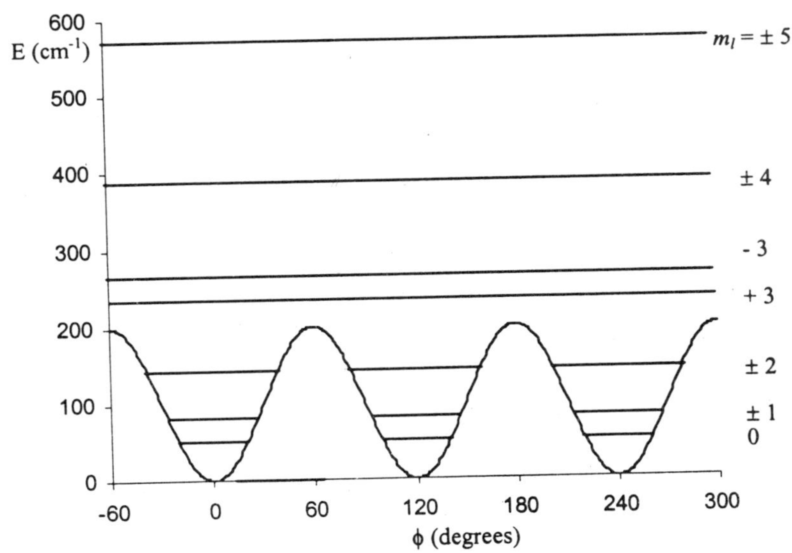

Figure 7-5. Internal rotation energy levels for a molecule with a barrier hindering rotation of 200 cm^{-1} and $E^{(0)} = m_l^2$ (17 cm^{-1}). Note that if the molecule is in the $m_l = 0$, ±1, or ±2 the molecule is classically locked into a particular configuration. Also note that the $m_l = \pm 3$ eigenstates are no longer degenerate.

$$V = \frac{V_3}{2}\left(1 - \cos 3\phi\right) \tag{7-25}$$

The term V_3 is the difference between the maxima and minima in the potential barrier. The Hamiltonian for the internal rotation is given as follows.

$$\hat{H} = -\frac{\hbar^2}{2I}\frac{d^2}{d\phi^2} + \frac{V_3}{2}\left(1 - \cos 3\phi\right) = \hat{H}^{(0)} + \hat{H}^{(1)} \tag{7-26}$$

Equation 3-6 gives the zero-order energy eigenvalues. Using Perturbation theory, the energy eigenvalue for the $m_l = 0$ state including a second-order correction summed up to $m_l = \pm 3$ unperturbed eigenstates is given as follows.

$$E_{m_l = 0} = \frac{V_3}{2} - \frac{I V_3^2}{36\hbar^2} \tag{7-27}$$

This analysis can be continued for other m_l eigenstates and what is observed is that the degeneracy is broken for some states. This is shown in Figure 7-4 and Figure 7-5 for the rotational barriers, V_3, of 100 and 200 cm^{-1} respectively. As can be seen in both figures, that if a molecule is in a low rotational m_l eigenstate, the molecule is classically locked into a particular configuration. Quantum mechanically, however, the molecule may be able to tunnel from one configuration to another depending on the height of the barrier relative to the energy of the eigenstate for the particular configuration.

7.3 INFRARED SPECTROSCOPY OF POLYATOMIC MOLECULES

The infrared spectra of polyatomic molecules involve vibrational transitions along with rotational transitions just like in diatomic molecules. However, especially in the case of low-resolution spectra of polyatomic molecules, the rotational fine structure is lost. The peaks in the infrared spectrum are assigned as a fundamental vibrational transition. Hence, the low-resolution spectrum represents a non-rotating or rotationally averaged model of the polyatomic molecule. The analysis that is done in this section assumes that the molecule is non-rotating and the frequencies obtained in a spectrum correspond to a vibrational transition.

Vibrations in polyatomic molecules can be quite complicated. The motion of a pair of nuclei many times cannot be isolated from the motion of other closely surrounding nuclei in the molecule. For this reason, the characteristic absorption of a particular functional group in a molecule is assigned to a range of characteristic frequencies where it can in general be found in an infrared spectrum (a few common functional groups are shown in Table 7-5). In order to analyze the vibrations of a molecule, it is helpful to determine the number of degrees of freedom available to vibration. A molecule with N atoms has a total of 3N degrees of freedom. Three degrees of freedom are in terms of translational motion. As discussed in the previous two sections, there are 2 degrees of rotational freedom for linear molecules and 3 degrees of rotational freedom for non-linear molecules. This leaves a total of 3N - 5 degrees of freedom for linear molecules and 3N - 6 degrees of freedom for non-linear molecules that do not depend on the position or orientation of the molecule in space. The remaining 3N - 5 and 3N - 6 degrees of freedom correspond to the ***internal coordinates*** of a molecule comprised of bond lengths and angles. As an example, a gas phase water

Table 7-5. The characteristic vibrational frequencies for some common functional groups are shown. The range of the frequencies is as a result of the dependence of what is bonded to these groups.

Bond	Group	Frequency (cm^{-1})
C=C	Alkene	1680-1600
	Aromatic	1600 and 1475
C≡C	Alkyne	2250-2100
C=O	Aldehyde	1740-1720
	Ketone	1725-1705
	Carboxylic Acid	1725-1700
	Ester	1750-1730
C-N	Amines	1350-1000
C=N	Imines and Oximes	1690-1640
C≡N	Nitriles	2260-2240
N=O	Nitro (R-NO_2)	1550 and 1350
C-F	Fluorides	1400-1000

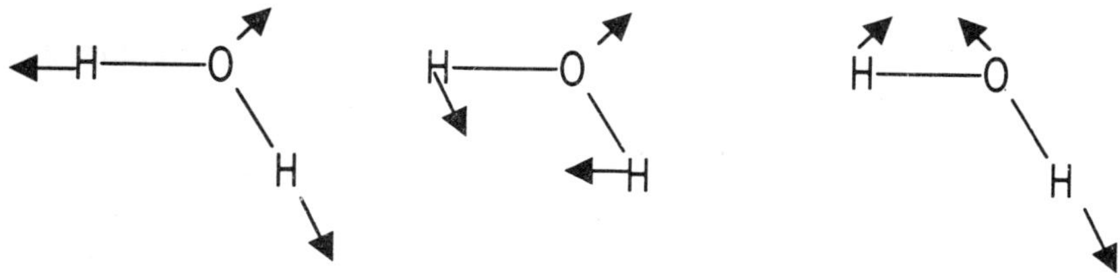

Figure 7-6. The normal modes of vibration for water are shown. (a) This corresponds to a symmetric stretch, ω_1 = 3,657.05 cm^{-1}. (b) This mode corresponds to a bend, ω_2 = 1,594.78 cm^{-1}. (c) This mode corresponds to an anti-symmetric stretch, ω_3 = 3,755.79 cm^{-1}.

molecule has a total of 9 degrees of freedom: 3 translational, 3 rotational (since it is non-linear), and 3 internal coordinates. The three internal coordinates correspond to the two O-H bond lengths and the bond angle.

The vibrations of a molecule set up a potential called a ***force field***. The force field is determined for a set of internal coordinates. If the force field is completely known, then the vibrations of the molecule are known. The force field can be obtained from absorptions in infrared spectra. The simplest force field model is the harmonic oscillator, and this will be used in the analysis here of the pure vibration of polyatomic molecules.

The Hamiltonian for a harmonic potential of a polyatomic molecule can be transformed from atomic displacement coordinates of the individual nuclei to ***normal coordinates*** whereby *separability* of the harmonic vibrations is achieved. The normal coordinates correspond to the actual vibrational modes that the molecule will undergo.

As an example of normal coordinates, consider the normal vibrational modes of water as shown in Figure 7-6. In each vibrational mode it is important that it not reflect either translation or rotation of the molecule in space. Note that for the symmetric stretch of water, the oxygen must also move out in the opposite direction of the hydrogen atoms though its motion is much less due its greater mass or else this mode would represent translation of the entire molecule in space. In the bending mode, again an arrow is needed for the movement of the oxygen or else this mode will represent rotation or net translation of the molecule.

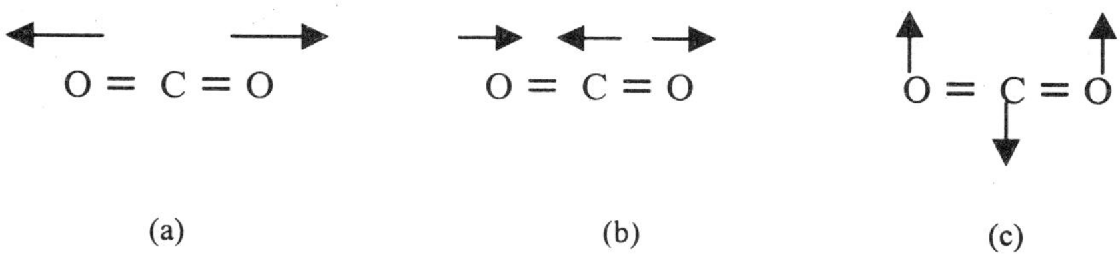

Figure 7-7. The normal modes of vibration for carbon dioxide are shown. (a) This corresponds to a symmetric stretch, $\omega_1 = 1388\ cm^{-1}$. (b) This mode corresponds to an anti-symmetric stretch, $\omega_2 = 2284\ cm^{-1}$. (c) This degenerate mode corresponds to a bend, $\omega_3 = 667\ cm^{-1}$. The other mode has the nuclei oscillating into and out of the plane of this page.

As another example, the modes of vibration can be analyzed for carbon dioxide, shown in Figure 7-7. Carbon dioxide is a linear molecule with a total of 9 degrees of freedom. There are 3 degrees of freedom for translation, and since it is linear, there are 2 rotational degrees of freedom. This leaves a total of 4 vibrational modes. The independent vibrational modes consist of symmetric and anti-symmetric stretches along with two degenerate bending modes, one in plane and the other out of plane.

The position of the nuclei is written as a set $\{q_1, q_2, \ldots, q_{3N-5}\}$ for linear molecules and $\{q_1, q_2, \ldots, q_{3N-6}\}$ for non-linear molecules that correspond to the particular normal modes of vibration. Each vibrational mode will have an effective mass m_q and effective force constant, k_q. The Hamiltonian for the harmonic vibration of polyatomic linear and non-linear molecules is given as follows.

$$\hat{H}^{linear} = \frac{1}{2}\sum_{i}^{3N-5}\left(\dot{q}_i^2 + \omega_i q_i^2\right) \tag{7-28a}$$

$$\hat{H}^{non-linear} = \frac{1}{2}\sum_{i}^{3N-6}\left(\dot{q}_i^2 + \omega_i q_i^2\right) \tag{7-28b}$$

Each vibrational mode represented by "i" is separable resulting in multiple vibrational Shroedinger equations mathematically equivalent to that for a diatomic molecule. The eigenfunctions and the vibrational energy eigenvalues will have the same form as for a diatomic molecule.

$$E^{linear}_{v_1,v_2,v_3,\ldots} = \sum_i^{3N-5} \left(v_i + \tfrac{1}{2}\right)\hbar\omega_i \tag{7-29a}$$

$$E^{non-linear}_{v_1,v_2,v_3,\ldots} = \sum_i^{3N-6} \left(v_i + \tfrac{1}{2}\right)\hbar\omega_i \tag{7-29b}$$

The values of v_i are 0, 1, 2, 3,..., and the ground-state vibrational energy will correspond to when all v_i 's are equal to zero. As an example, the harmonic vibrational energy for a water molecule is specified by three quantum numbers. The ground vibrational state for water is (0, 0, 0). An excited vibrational state for water will correspond to one or more vibrational modes at some value above zero such as (0, 1, 0) or (1, 1, 2) and so forth. Some of the low-lying harmonic vibrational levels for water are shown in Figure 7-8. Note that as the vibrational energy increases, the number of vibrational states in some small increment of energy increases. The number of states within a small increment of energy is called the ***density of states***. The density of vibrational states for polyatomic molecules increases with increasing energy.

The selection rules for vibrational transitions of polyatomic molecules are the same as for diatomic molecules. The selection rule results for diatomic molecules can be applied to the harmonic model for vibrations of polyatomic molecules because the separation of variables achieved in the result states that each normal mode for vibration is regarded as a 1-dimensional oscillator. For the harmonic model, it was found that the quantum number changes by one. Also the dipole moment of the molecule must also change in the course of the transition in order for it to be an allowed transition. A molecule such as O_2 does not have an allowed fundamental infrared transition whereas HCl does.

In order to determine whether a particular vibrational mode of a polyatomic molecule will be active in the infrared, the molecule's dipole must be assessed to see if it changes in a normal mode vibration. Classically the dipole moment of a molecule is determined as follows.

$$\mu_x = \sum_i q_i x_i \tag{7-30a}$$

$$\mu_y = \sum_i q_i x_y \tag{7-30b}$$

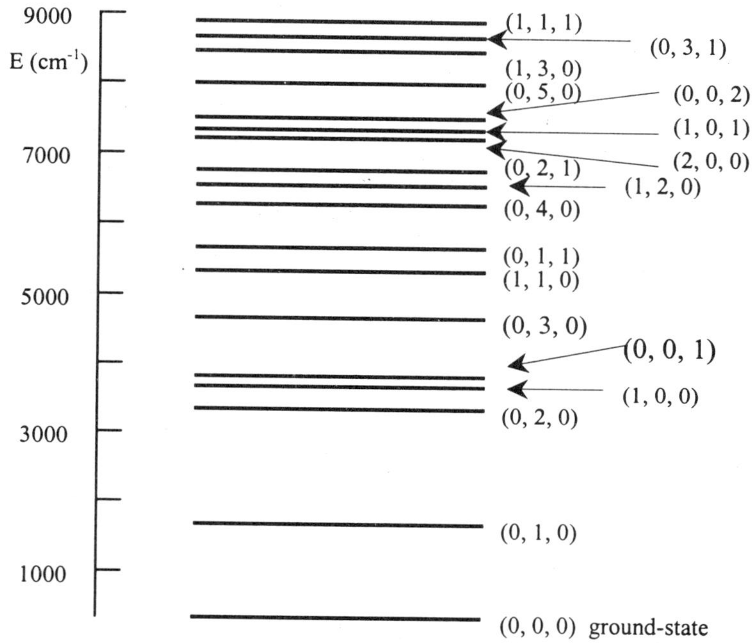

Figure 7-8. The low-lying harmonic vibrational energy levels for a gas phase water molecule are shown. Note that as the energy increases, the density of states increases.

$$\mu_z = \sum_i q_i x_z \qquad (7\text{-}30c)$$

The q_i in Equations 7-30a-c refers to the charge of the i-th particle. Following a simple scheme and Equations 7-30a-c, a qualitative determination of whether the dipole moment of a molecule is changing in the course of absorption can be made. First, all atoms are considered to have a partial charge, and atoms that are chemically equivalent are assumed to have the same charge. The displacement of the atoms in a normal mode of vibration is then considered to determine if there is a change in the dipole moment of the molecule.

As an example, consider carbon dioxide where the normal modes of vibration are shown in Figure 7-7. The oxygen atoms are more electronegative than the carbon atoms; hence, the oxygen atoms are assigned a charge of -δ. The carbon must then have a charge of +2δ in order for the molecule to be neutral. The carbon atom is placed at the origin of the coordinate system. In the symmetric stretch mode (see Figure 7-7a), both of the oxygen atoms oscillate in-line with the carbon atom. Any change in the dipole moment of the molecule that occurs as a result of one oxygen atom oscillating is negated by the oscillation of the other oxygen atom. Consequently, the symmetric stretch of carbon dioxide will not result in a change in the dipole moment of the molecule and is a forbidden transition in infrared spectroscopy. In the asymmetric stretch (see Figure 7-7b), the oxygen atoms are oscillating counter to one another so the change in dipole moment that each oxygen atom contributes is not negated. The dipole moment of the molecule does change in this vibrational mode and it results in an allowed transition.

In the case of diatomic molecules, the strongest absorption peak in an infrared spectrum is $\Delta\nu = 1$ ($\nu = 0 \rightarrow \nu = 1$). Other transitions are possible due to anharmonicity effects; however, the $\Delta\nu = 1$ transition is expected to be the most dominant in the spectrum. For polyatomic molecules, the selection rule for harmonic vibrational transitions is also $\Delta\nu_i = 1$. This means that one vibrational mode undergoes a transition while the other vibrational modes do not change. Other vibrational transitions may become allowed due to anharmonicity; however, it is expected that the harmonic selection rule of $\Delta\nu_i = 1$ while $\Delta\nu$ for all the other modes as zero will be the strongest absorptions in the spectrum. These types of transitions, when it originates from the ground-state, are called ***fundamental transitions***. When $\Delta\nu = 1$ but does not originate from the ground-state, it is called a ***hot band***. When $\Delta\nu > 1$, these transitions are called ***overtone transitions***. When more than one vibrational mode undergoes a transition it is called a ***combination transition***.

As an example, consider the possible vibrational transitions for a sample of low-density gas phase water (no hydrogen bonding) in an infrared spectrum. The vibrational quantum numbers is represented as (ν_1, ν_2, ν_3) where ν_1 represents the symmetric stretch, ν_2 represents the asymmetric stretch, and ν_3 represents the bending mode. The following list represents some possible transitions and there designation.

$(0, 0, 0) \rightarrow (1, 0, 0)$	fundamental
$(0, 0, 0) \rightarrow (0, 1, 0)$	fundamental

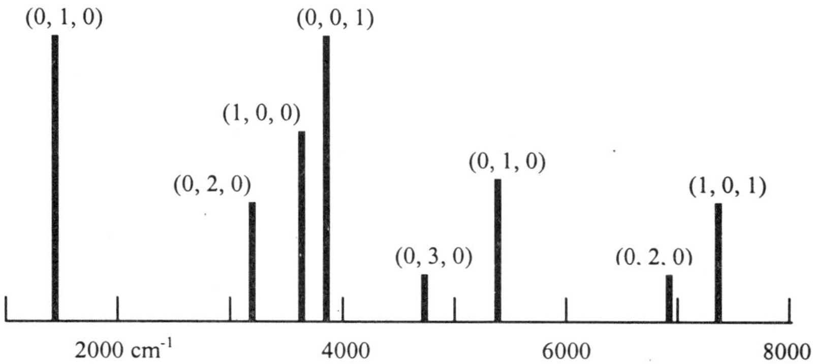

Figure 7-9. An idealized infrared spectrum of low-density gas phase water where there is no hydrogen bonding. The heights of the bars reflect the relative intensities of the anticipated transitions. The ground-state in each transition is assumed to be in the ground-state (0, 0, 0), and the state listed above the peak is the final state.

(0, 0, 0) → (0, 2, 0)	overtone
(1, 0, 0) → (2, 0, 0)	hotband
(0, 0, 0) → (1, 0, 1)	combination

An idealized infrared spectrum of water is shown in Figure 7-9. Note that the fundamental transitions are the most intense absorptions in the spectrum.

Another type of transition that may be observed in an infrared spectrum is one where the initial state is not in the ground vibrational state. This type of peak in the infrared spectrum is called a ***hot band***. Hot bands may appear in a spectrum either very close or essentially on top of other fundamental transition peaks. As an example, the hot band transition of (0, 1, 0) → (0, 2, 0) is expected to fall in the same region as the fundamental transition of (0, 0, 0) → (0, 1, 0). The presence and identification of hot bands in an infrared spectrum can be determined by obtaining the spectrum at a higher temperature. The peaks of transitions involving excited initial states will grow, helping to discern these peaks from the fundamental transitions.

PROBLEMS AND EXERCISES

7.1) Based on the normal modes of vibration for CO_2 shown in Figure 7-7, determine which modes are infrared active.

7.2) Based on the following rotational constants for the linear molecule NNO, determine the substitution structure of the molecule.

$^{14}N^{14}N^{16}O$	12,561.7 MHz
$^{15}N^{14}N^{16}O$	12,137.3 MHz
$^{14}N^{14}N^{18}O$	11,859.1 MHz

7.3) How many normal vibrational modes will the following molecules have? H_2O_2, C_2H_2 (acetylene), C_2H_4 (ethylene), C_2H_3Cl (ethylene chloride), C_6H_6 (benzene), C_6H_5Cl (chlorobenzene)

7.4) The rotational constant for CO_2 was measured to be B = 0.3937 cm^{-1}. Calculate the CO bond length.

7.5) Based on the rotational constants and the bond lengths for formaldehyde (CH_2O), determine the bond angle.

A = 282.106 GHz B = 38.834 GHz C = 34.004 GHz
C-H: 0.107 nm C=O: 0.122 nm

7.6) Classify the following molecules as asymmetric top, symmetric top, or spherical top: CH_4, CH_3Cl, CH_2Cl_2, C_2H_2, benzene, SF_6.

7.7) Using second-order Perturbation Theory, confirm Equation 7-27 summing up to the $m_l = \pm 5$ rotational states.

7.8) Make a sketch like that in Figures 7-4 and 7-5 of the rotational levels for a molecule with an internal degree of rotation whereby V_3 = 400 cm^{-1}.

7.9) Find the center of mass and moments of inertia for $^2H_2{}^{16}O$ and $^1H_2{}^{18}O$ assuming that the bond lengths are the same as for $^1H_2{}^{16}O$.

Chapter 8

Atomic Structure and Spectra

In this chapter, the electronic structure of atoms will be analyzed. The simplest atom, the hydrogen atom, is an important model problem for the electronic structure of atoms. Sadly, the hydrogen atom, and other one-electron systems, is the last model problem for which an exact solution of the Schroedinger equation can be obtained. Heavier atoms, and subsequently molecules in the next chapter, will require some degree of approximation in order to solve the Schroedinger equation. The results of the Schroedinger equation for a hydrogen atom will be used as a basis to solve for the electronic structure of heavier atoms along with obtaining an understanding of electronic spectra of atoms.

8.1 One-Electron Systems

The system described here is a two-body system (such as a hydrogen atom or a He^+ ion): a positively charged nucleus and a negatively charged electron separated by a distance r. There is a potential between the two bodies along the radial coordinate r; however, the system is free to rotate about the angular coordinates θ and ϕ. Mechanically this is similar to the vibration/rotation of a diatomic molecule resulting in the same Schroedinger equation in terms of the general expression for the potential V(r) (see Section 6.2).

$$\hat{H}(r,\theta,\phi)\psi(r,\theta,\phi) = E\psi(r,\theta,\phi)$$

$$-\frac{\hbar^2}{2\mu}\nabla^2\psi(r,\theta,\phi) + V(r)\psi(r,\theta,\phi) = E\psi(r,\theta,\phi)$$

$$-\frac{\hbar^2}{2\mu r}\left(\frac{\partial^2(r\cdot\psi)}{\partial r^2}\right) - \frac{\hbar^2}{2\mu r^2}\Lambda^2\psi + V(r)\psi = E\psi \qquad (8\text{-}1)$$

The reduced mass, μ, in this system is determined from the mass of an electron, m_e, and the atomic mass of the nucleus, m_N. However, the mass of a proton and neutron is much more massive than an electron (1836.5 and 1838.7 times more massive respectively); hence, the reduced mass for the system can be taken as the mass of an electron.

$$\mu = \frac{m_e m_N}{m_e + m_N} \cong m_e \qquad (8\text{-}2)$$

Since the radial and angular components are separable, the wavefunction will be a product of the angular function and a radial function, $R_{nl}(\mathrm{r})$. The system is free to rotate about the θ and ϕ axes as in the Particle-on-a-Sphere model problem; hence, the angular wavefunctions are the spherical harmonics, $Y_{l,m}(\theta, \phi)$.

$$\psi(r,\theta,\phi) = Y_{lm}(\theta,\phi)R_{nl}(r) \qquad (8\text{-}3)$$

To no surprise, substitution of Equation 8-3 into Equation 8-1 along with operation of Λ^2 on $Y_{lm}(\theta, \phi)$ and subsequent cancellation of $Y_{lm}(\theta, \phi)$ results in the same two-body radial Schroedinger equation as previously obtained for the vibration/rotation of diatomic molecules with a general expression for the potential V(r) (see Equation 6-10).

$$-\frac{\hbar^2}{2m_e r}\left(\frac{\partial^2(r\cdot R_{nl})}{\partial r^2}\right) + \frac{\hbar^2 l(l+1)}{2m_e r^2}R_{nl} + V(r)R_{nl} = ER_{nl} \qquad (8\text{-}4)$$

The potential V(r) along the radial coordinate is Coulombic. The charge of the nucleus is +Ze where Z is the atomic number and e is the elementary charge (e = 1.602177 x 10^{-19}C). The charge of the electron is equal to -e.

$$V(r) = -\frac{Ze^2}{4\pi\varepsilon_0 r} \tag{8-5}$$

The term ε_0 is the vacuum permittivity constant which in SI units is equal to 8.85419 x 10^{-12} J^{-1} C^2 m^{-1}. Substitution of Equation 8-5 into Equation 8-4 results in the following radial Schroedinger equation for a one-electron system.

$$-\frac{\hbar^2}{2m_e r}\left(\frac{\partial^2(r \cdot R_{nl})}{\partial r^2}\right) + \frac{\hbar^2 l(l+1)}{2m_e r^2} R_{nl} - \frac{Ze^2}{4\pi\varepsilon_0 r} R_{nl} = ER_{nl} \tag{8-6}$$

In the case of the vibration/rotation of a diatomic molecule, the $l(l + 1)$ term in the radial Schroedinger equation is approximated via a power series expansion (see Equation 6-18). This approximation is sufficient for vibration/rotation of diatomic molecules because the distance of separation of the two nuclei does not vary greatly between rotational states. In the case of electronic states however, the separation of the electron and the nucleus varies widely between states and a power series expansion is inappropriate. Fortunately, the solution to Equation 8-6 is well known. There are an infinite number of solutions for each value of l and each one is designated by a quantum number n. Each state is called an ***atomic orbital (AO)***. The quantum numbers that distinguish the possible states is given as follows.

n = 1, 2, 3, ...

l = 0, 1, 2, 3, ..., n - 1

m_l = $-l$, $-l$ + 1, ..., l - 1, l

Before describing the radial functions $R_{nl}(r)$, it is convenient to change the units. In terms of SI units, the energy eigenvalues obtained from the solution of Equation 8-6 will be in Joules. SI units are not convenient for systems at the atomic or molecular scale. Most quantum chemists report the results of their calculations in ***atomic units***. In atomic units, the unit of mass is in terms of the mass of an electron instead of kg (i.e. the mass expressed as a dimensionless factor of the mass of the system to that of an electron: μ/m_e). The unit for angular momentum is a dimensionless factor in terms of $\hbar$ rather than kg m^2 s^{-1}. The radial coordinate is expressed as a ratio of the distance of separation between the electron and nucleus to the Bohr radius, a_0. The Bohr radius is the distance of separation between the proton and electron in the ground-state of a hydrogen atom obtained from classical mechanics (a_0 = 5.29177 x 10^{-11} m). Charge is expressed in terms of a dimensionless ratio of the charge q to that of the unit charge e with the constant $4\pi\varepsilon_0$ included. The net result of atomic units is to make the quantities $\hbar$, m_e, and the charge q (with the combined constants) equal to 1. The Hamiltonian for a one-electron system with an atomic number Z in atomic units is as follows.

$$\hat{H}(r,\theta,\phi) = -\frac{1}{2}\nabla^2 - \frac{Z}{r} \tag{8-7}$$

The radial Schroedinger equation for a one-electron system in atomic units is reduced to the following expression.

$$-\frac{1}{2r}\left(\frac{\partial^2(r \cdot R_{nl})}{\partial r^2}\right) + \frac{l(l+1)}{2r^2}R_{nl} - \frac{ZR_{nl}}{r} = ER_{nl} \tag{8-8}$$

The energy eigenvalues in atomic units obtained from the solution of Equation 8-8 is expressed in ***hartrees (h)***.

1 hartree = 27.212 eV = 4.3599 x 10^{-18} J

Another type of atomic unit for energy that is occasionally used by chemists to report their computational results is in ***rydbergs***.

1 rydberg = ½ hartree

The radial functions $R_{nl}(r)$ that satisfy Equation 8-8 are constructed from the ***Leguerre polynomials***. The Leguerre polynomial of an order k can be generated from the following expression.

$$L_k(z) = e^z \frac{d^k}{dz^k}\left(z^k e^{-z}\right) \tag{8-9}$$

The associated Leguerre polynomials can be generated from the following expression.

$$L_k^j(z) = \frac{d^j}{dz^j} L_k(z) \tag{8-10}$$

The normalized radial functions $R_{nl}(r)$ in terms of the Leguerre polynomials are expressed as follows.

$$R_{nl}(r) = \sqrt{\frac{(n-1-1)!}{2n[(n+1)!]^3}} e^{-\rho/2} \rho^l L_{n+1}^{2l+1}(\rho) \tag{8-11}$$

$$\rho = \frac{2Z}{n} r \qquad \text{(atomic units)} \tag{8-12}$$

Table 8-1 lists the explicit form for a number of radial functions in atomic units.

The energy eigenvalues in hartrees for a one-electron system following substitution of the radial functions into Equation 8-8 are as follows.

$$E_n(hartrees) = -\frac{Z^2}{2n^2} \tag{8-13}$$

As can be seen by Equation 8-13, the energy eigenvalues depend only on the n quantum number. Note that the energy is negative indicating that the electron remains around the nucleus by Coulombic attraction. For the ground-state of hydrogen (Z = 1), the energy is -½ hartree or -1 rydberg. For n = 2, the energy for hydrogen is -1/8 hartree. As the value of n increases, the energy of the system approaches zero and the energy difference between levels becomes smaller reaching a continuum of states. The limit that n approaches infinity and the energy approaches zero corresponds to ionization. The electron at that point is completely separated from the nucleus. Since the energy of the one-electron system depends only on the quantum number n, the energy levels are degenerate. The degeneracy of each level is n^2.

Table 8-1. The normalized one-electron radial functions, R_{nl}, in atomic units are shown below where Z is the atomic number and

$$\rho = \frac{2Z}{n} r \, .$$

n	*l*	$R_{nl}(r)$
1	0	$R_{10}(r) = Z^{\frac{3}{2}} 2e^{-\rho/2}$
2	0	$R_{21}(r) = \frac{Z^{\frac{3}{2}}}{2\sqrt{2}}(2-\rho)e^{-\rho/2}$
	1	$R_{21}(r) = \frac{Z^{\frac{3}{2}}}{2\sqrt{6}} \rho e^{-\rho/2}$
3	0	$R_{30}(r) = \frac{Z^{\frac{3}{2}}}{9\sqrt{3}}(6-6\rho+\rho^2)e^{-\rho/2}$
	1	$R_{31}(r) = \frac{Z^{\frac{3}{2}}}{9\sqrt{6}}(4\rho-\rho^2)e^{-\rho/2}$
	2	$R_{32}(r) = \frac{Z^{\frac{3}{2}}}{9\sqrt{30}} \rho^2 e^{-\rho/2}$
4	0	$R_{40}(r) = \frac{Z^{\frac{3}{2}}}{96}(24-36\rho+12\rho^2-\rho^3)e^{-\rho/2}$
	1	$R_{41}(r) = \frac{Z^{\frac{3}{2}}}{32\sqrt{15}}(20\rho-10\rho^2+\rho^3)e^{-\rho/2}$
	2	$R_{42}(r) = \frac{Z^{\frac{3}{2}}}{96\sqrt{5}}(6\rho^2-\rho^3)e^{-\rho/2}$
	3	$R_{43}(r) = \frac{Z^{\frac{3}{2}}}{96\sqrt{35}} \rho^3 e^{-\rho/2}$

Setting the following integral equal to one normalizes the wavefunctions.

$$\int_0^{2\pi}\int_0^{\pi}\int_0^{\infty}\psi^*_{nlm}\psi_{nlm}r^2\,dr\sin\theta d\theta d\phi = 1 \tag{8-14}$$

It is usual to normalize the angular and radial parts separately. The radial functions listed in Table 8-1 are normalized along the r coordinate, and the spherical harmonic wavefunctions in Table 3-1 are also normalized.

Example 8-1

Problem: Confirm that the n = 2, l = 1, m_l = -1 ($\psi_{21\text{-}1}$) wavefunction is normalized.

Solution: The $\psi_{21\text{-}1}$ is constructed from the radial function R_{21} in Table 8-1 and the $Y_{1\text{-}1}$ spherical harmonic function in Table 3-1.

$$\psi_{21-1} = \left(\left(\frac{Z^{\frac{3}{2}}}{2\sqrt{6}}\right)Zre^{-Zr/2}\right)\left(\tfrac{1}{2}\sqrt{\tfrac{3}{2\pi}}\sin\theta e^{-i\phi}\right)$$

Substitution of the wavefunction into Equation 8-14 results in the following integral to be solved.

$$\frac{3Z^5}{96(2\pi)}\left(\int_0^{\infty}r^4e^{-Zr}dr\right)\left(\int_0^{2\pi}\sin^3\theta d\theta\right)\left(\int_0^{2\pi}d\phi\right)$$

$$= \frac{3Z^5}{96(2\pi)}\left(\frac{24}{Z^5}\right)\left(\frac{4}{3}\right)(2\pi) = 1$$

The integral is equal to one confirming that the $\psi_{21\text{-}1}$ is normalized.

The wavefunctions for a one-electron system and the associated radial functions can now be examined in detail. Below is a list of the first two n-states of a one-electron system.

The 1s function: n = 1, l = 0, m_l = 0; $\psi_{100} = R_{10}Y_{00} = \frac{Z^{\frac{3}{2}}}{\sqrt{\pi}}e^{-\rho/2}$

The radial function is shown in Figure 8-1, and the radial distribution function (r^2R) for this eigenstate is shown in Figure 8-2. The radial distribution peaks at r = a_0, the Bohr radius. The radial distribution of the electron is not confined to a limited sphere around the nucleus but rather it dies away smoothly at large values of r. Since l = 0, the orbital angular momentum of the electron around the nucleus is zero. This defeats the notion that the electron "orbits" around the nucleus.

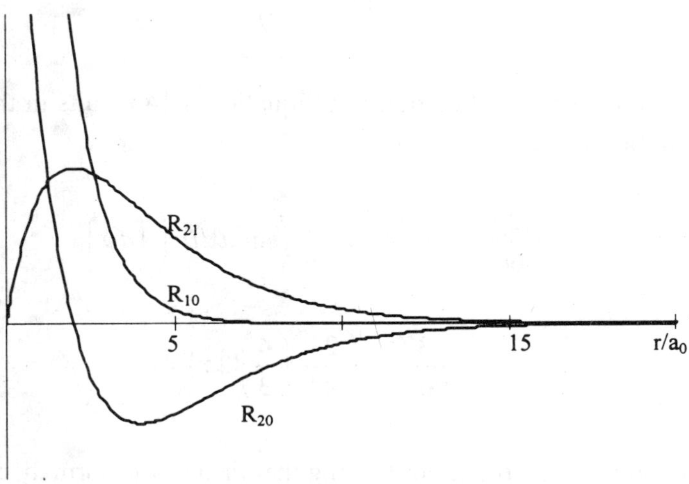

Figure 8-1. The radial functions for the first two levels of hydrogen are shown. The R_{10} and R_{20} functions correspond to 1s and 2s orbitals respectively. The R_{21} radial function corresponds to a 2p orbital.

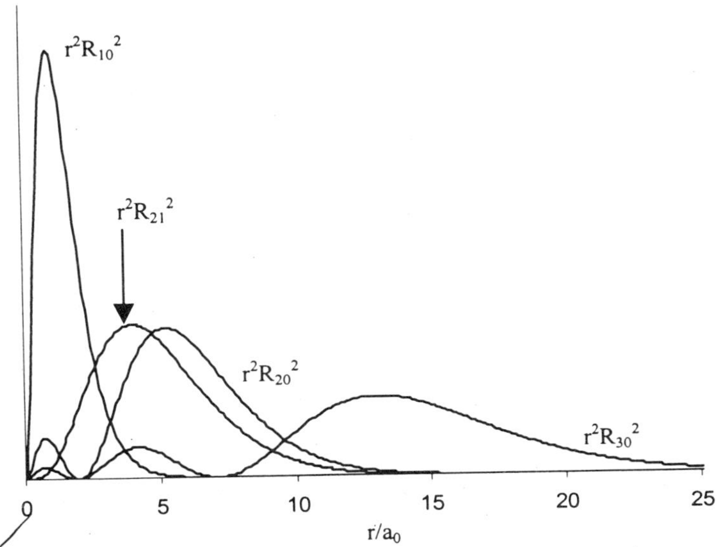

Figure 8-2. The radial probability functions, r^2R^2, of a hydrogen atom for the first two energy levels are shown. Note that the probability functions decay to zero as r approaches infinity.

The 2s function: n = 2, l = 0, m_l = 0; $\psi_{200} = R_{20}Y_{00} = \frac{Z^{\frac{3}{2}}}{4\sqrt{2\pi}}(2-\rho)e^{-\rho/2}$

The radial function is shown in Figure 8-1, and the radial distribution function (r^2R) for this eigenstate is shown in Figure 8-2. The radial function has one node at r = $2a_0$ representing a point of zero probability.

The 2p functions: n = 2, l = 1, m_l = 0, ±1.

for m_l = 0: $$2p_z = \psi_{210} = \frac{Z^{\frac{3}{2}}}{4\sqrt{2\pi}}\rho\cos\theta e^{-\rho/2}$$

This p-orbital is designated as p_z because z = rcosθ (see expression for ρ in Equation 8-12). The function has a maximum value at θ equal to 0° and 90° (at θ = 90°, the function is in the x-y plane). The function is negative below the x-y plane and positive above the x-y plane.

for $m_l = +1$:
$$2p_1 = \psi_{211} = -\frac{Z^{\frac{3}{2}}}{8\sqrt{\pi}}\rho \sin\theta e^{i\phi} e^{-\rho/2}$$

for $m_l = -1$:
$$2p_{-1} = \psi_{21-1} = \frac{Z^{\frac{3}{2}}}{8\sqrt{\pi}}\rho \sin\theta e^{-i\phi} e^{-\rho/2}$$

Spatially, the $2p_1$ and $2p_{-1}$ functions are the same since

$$\psi^*_{21\pm1}\psi_{21\pm1} = \frac{Z^3}{64\pi}\rho^2 e^{-\rho}\sin^2\theta$$

for each function. The density of each function is equal to zero along the z-axis. Operation of the angular momentum operator along the z-axis,

$$\hat{L}_z = \frac{\hbar}{i}\frac{\partial}{\partial\phi},$$

results in equal but opposite orbital angular momentum of $-\hbar$ and $\hbar$ for $2p_{-1}$ and $2p_1$ respectively. This would indicate that the electron is circulating in opposite directions about the z-axis. Since the $2p_1$ and $2p_{-1}$ states are degenerate (assuming that there is no external magnetic field), any linear combination of the two functions can be used. The two linear combinations that are most often used are as follows.

$$2p_x = \frac{1}{\sqrt{2}}(\psi_{211} - \psi_{21-1}) = \frac{\sqrt{2}Z^{\frac{3}{2}}}{8\sqrt{\pi}}\rho e^{-\rho/2}\sin\theta\cos\phi$$

$$2p_y = \frac{i}{\sqrt{2}}(\psi_{21} + \psi_{21-1}) = \frac{\sqrt{2}Z^{\frac{3}{2}}}{8\sqrt{\pi}}\rho e^{-\rho/2}\sin\theta\sin\phi$$

Since x = rsinθcosϕ and y = rsinθsinϕ, the wavefunctions are along the x and y-axes.

$$2p_x = \frac{\sqrt{2}Z^{\frac{5}{2}}}{8\sqrt{\pi}} r\sin\theta\cos\phi e^{-\rho/2} = \frac{\sqrt{2}Z^{\frac{5}{2}}}{8\sqrt{\pi}} xe^{-\rho/2}$$

$$2p_y = \frac{\sqrt{2}Z^{\frac{5}{2}}}{8\sqrt{\pi}} r\sin\theta\sin\phi e^{-\rho/2} = \frac{\sqrt{2}Z^{\frac{5}{2}}}{8\sqrt{\pi}} ye^{-\rho/2}$$

Example 8-2

Problem: Determine the average radial position, r, of the electron in a 1s orbital of a hydrogen atom and a He^+ ion.

Solution: Since the radial and angular parts of the wavefunction are separable, only the radial function for a 1s orbital is needed, R_{10}. This radial function is obtained from Table 8-1.

$$R_{10}(r) = Z^{\frac{3}{2}} 2e^{-\rho/2} = Z^{\frac{3}{2}} 2e^{-Zr}$$

$$<r>_{1s} = \langle R_{10} | r | R_{10} \rangle = 4Z^3 \int_0^\infty r^3 e^{-2Zr} dr = 4Z^3 \left(\frac{6}{16Z^4} \right) = \frac{1.5}{Z}$$

The average radial position of the electron in a hydrogen (Z = 1) 1s orbital is $1.5a_0$ (1.5 x 52.9177 pm = 79. pm). For a He^+ ion (Z = 2), the average radial distance is equal to $0.75a_0$. As expected, increasing the charge of the nucleus brings the electron on average closer to the nucleus.

The role of angular momentum in the orbitals of a one-electron system can be analyzed by rewriting the radial Schroedinger equation (Equation 8-8) in terms of an effective potential $V^{eff}(r)$.

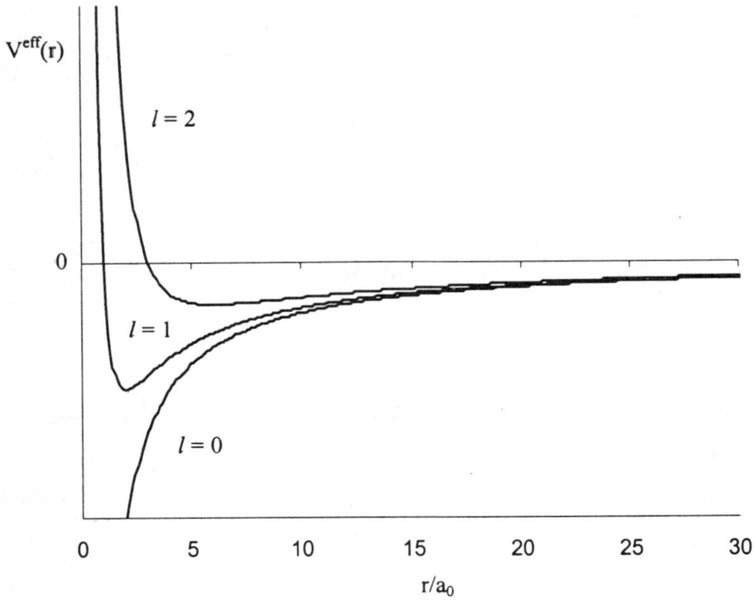

Figure 8-3. The effective potential (Equation 8-15) for a hydrogen atom is shown for an $l = 0$, $l = 1$, and $l = 2$ orbital. For $l = 0$, there is no repulsive potential and the attractive potential continues to increase (more negative) at small radial distances, r. For $l \neq 0$, the repulsive potential dominates at small radial distances.

$$-\frac{1}{2}\left(\frac{\partial^2 (r \cdot R_{nl})}{\partial r^2}\right) + V^{eff}(r) R_{nl} = E R_{nl}$$

$$V^{eff}(r) = \frac{l(l+1)}{2r^2} - \frac{Z}{r} \tag{8-15}$$

The first term of the effective potential in Equation 8-15 represents the repulsive centrifugal potential energy whereas the second term is the attractive Coulombic potential energy. When $l = 0$ (an s-orbital), there is no repulsive potential energy, only attractive Coulombic potential energy (see Figure 8-3). This results in a non-zero probability of finding the electron at the nucleus. This conclusion is reflected in the R_{n0} radial functions that have non-zero values at r = 0 (see Figure 8-1). For orbitals where $l \neq 0$, the repulsive centrifugal potential energy is enough to overcome the attractive Coulombic potential at short distances forbidding the electron to be found at the nucleus (see Figure 8-3). This is reflected in the radial functions with non-zero *l* values as they all have a node at r = 0.

Since the radial functions for the atomic orbitals decay to zero at large values of r, it is convenient to discuss the shapes of the orbitals in terms of a percentage of the total ***electron density***. The probability of finding an electron in a sphere of radius R can be found by solving the following integral.

$$P(r) = \int_0^{2\pi}\int_0^{\pi}\int_0^{R} \psi^*_{nlm}\psi_{nlm} r^2 dr \sin\theta d\theta d\phi \tag{8-16}$$

An electron density for a limited region of space is then determined by selecting some arbitrary value for P(r) such as 90%. The picture for each orbital is constructed from such a computation resulting in the familiar pictures such as a sphere for the s-orbitals and a dumbbell shape for the p-orbitals, and so on.

Example 8-3

Problem: What is the most probable point of finding an electron in the $2p_z$ (ψ_{210}) orbital? What is the probability of finding an electron within a sphere of radius R centered on the nucleus for the $2p_z$ orbital?

Solution: The first question is answered by finding the maximum value of ψ^2_{210}.

$$\psi_{210} = \frac{Z^{\frac{3}{2}}}{4\sqrt{2\pi}} \rho \cos\theta e^{-\rho/2}$$

$$\psi^2_{210} = \frac{Z^3}{32\pi} \rho^2 \cos^2\theta e^{-\rho}$$

The maximum value of ψ_{210}^2 will occur at when $\cos^2\theta = 1$. This occurs when θ is equal to 0 and π. Taking the first derivative with respect to r (in terms of ρ) and setting it equal to zero then determines the maximum of the function. The normalization constant has been cancelled from the expression.

$$\frac{d}{d\rho}\left(\rho^2 e^{-\rho}\right) = 0$$

$$2\rho e^{-\rho} - \rho^2 e^{-\rho} = 0; \qquad \rho = 2; \qquad r = \frac{2}{Z} \quad \text{(atomic units)}$$

The most probable point of finding the electron in a $2p_z$ orbital is at θ = 0 or π and r = (2/Z) in atomic units along the positive and negative z-axis.

The second question is determined by solving the integral in Equation 8-16 for a $2p_z$ orbital.

$$P(R) = \frac{Z^3}{32\pi}\int_0^{2\pi}\int_0^{\pi}\int_0^{R} \rho^2 \cos^2\theta e^{-\rho} r^2 dr \sin\theta d\theta d\phi$$

$$= \frac{Z^3}{32\pi}\left\{\frac{1}{Z^3}\int_0^{ZR}\rho^4 e^{-\rho} d\rho\right\}\left\{\int_0^{\pi}\cos^2\theta\sin\theta d\theta\right\}\left\{\int_0^{2\pi} d\phi\right\}$$

$$= \frac{1}{32\pi}\left\{24 - e^{-ZR}\left(Z^4R^4 + 4Z^3R^3 + 12Z^2R^2 + 24ZR + 24\right)\right\}\left\{\frac{2}{3}\right\}\{2\pi\}$$

$$= 1 - e^{-ZR}\left(\frac{Z^4R^4}{24} + \frac{Z^3R^3}{6} + \frac{Z^2R^2}{2} + ZR + 1\right)$$

When R = 2/Z, the probability is as follows.

$$P(2/Z) = 1 - e^{-2}(6.83) = 0.08$$

This result indicates that a significant probability of the electron extends beyond the most probable point along the radial coordinate.

Point of Further Understanding

From the standpoint of classical mechanics, the 1s orbital of a one-electron system poses somewhat of a paradox. The angular momentum of the electron is zero since $l = 0$, as pointed out previously. This would indicate that the electron is not "orbiting" around the nucleus. In addition, the radial wavefunction R_{10} indicates that the there is a non-zero probability of the electron being at the nucleus. Based on the premises of quantum mechanics, explain why the electron in a 1s orbital does not simply collapse into the nucleus canceling the charges?

8.2 THE HELIUM ATOM

The helium atom consists of a system with two electrons around a nucleus. This model can be applied to any two-electron system with an atomic number Z including H^-, Li^+, and Be^{2+}. The Hamiltonian includes kinetic energy operators for the two electrons, the Coulombic repulsion potential between the electrons, and a Coulombic attraction between each electron and the nucleus. This is shown schematically in Figure 8-4.

$$\hat{H} = -\frac{1}{2}\nabla_1^2 - \frac{1}{2}\nabla_2^2 - \frac{Z}{r_1} - \frac{Z}{r_2} + \frac{1}{r_{12}} \tag{8-17}$$

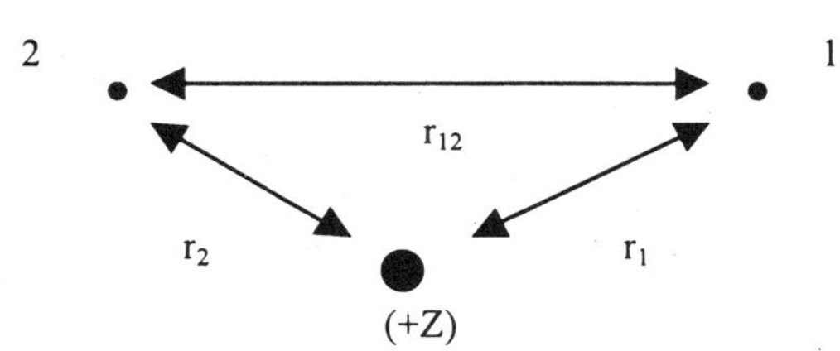

Figure 8-4. A schematic of a two-electron system is shown. Each electron has three coordinates to specify their position. The term r_{12} is the distance of separation between the electrons.

The term r_{12} is the distance of separation between the two electrons.

The Schroedinger equation involves six independent variables. Three coordinates specify each electron. The wavefunction is expressed as a function of these six coordinates.

$$\psi = f(r_1, \theta_1, \phi_1, r_2, \theta_2, \phi_2) \tag{8-18}$$

The coordinate r_{12} is in terms of the coordinates of each electron. In Cartesian coordinates, it is given as follows.

$$r_{12} = \sqrt{(x_1 - x_2)^2 + (y_1 - y_2)^2 + (z_1 - z_2)^2} \tag{8-19}$$

The r_{12} term cannot be separated into coordinates for either electron making the Hamiltonian for the two-electron system inseparable. In order to solve the Schroedinger equation, an approximation technique is needed.

A natural choice of approximation techniques to use for this system is perturbation theory. The Hamiltonian in Equation 8-17 can be divided into the unperturbed Hamiltonian, $\hat{H}^{(0)}$, consisting of the two hydrogen-like Hamiltonians, and the first-order perturbing Hamiltonian, $\hat{H}^{(1)}$, that consists of the electron-electron repulsion term $1/r_{12}$.

$$\hat{H}^{(0)} = \hat{H}_1 + \hat{H}_2 \tag{8-20}$$

$$\hat{H}_1 = -\frac{1}{2}\nabla_1^2 - \frac{Z}{r_1} \tag{8-21}$$

$$\hat{H}_2 = -\frac{1}{2}\nabla_2^2 - \frac{Z}{r_2} \tag{8-22}$$

$$\hat{H}^{(1)} = \frac{1}{r_{12}} \tag{8-23}$$

In the unperturbed system, the two electrons do not interact with one another. This means that the electrons are independent, and their motions are separable. The unperturbed wavefunction consists of a product of two hydrogen-like eigenfunctions.

$$\psi^{(0)}(r_1,\theta_1,\phi_1,r_2,\theta_2,\phi_2)=\psi_1(r_1,\theta_1,\phi_1)\psi_2(r_2,\theta_2,\phi_2) \tag{8-24}$$

The unperturbed wavefunction in Equation 8-24 can now be applied to the unperturbed Hamiltonian in Equation 8-20. This yields the unperturbed energy that consists of nothing more than a sum of two hydrogen-like eigenvalues as found previously in Equation 8-13.

$$\begin{aligned}\hat{H}^{(0)}\psi^{(0)} &= (\hat{H}_1+\hat{H}_2)\psi_1(r_1,\theta_1,\phi_1)\psi_2(r_2,\theta_2,\phi_2)\\ &= (E_1+E_2)\psi_1(r_1,\theta_1,\phi_1)\psi_2(r_2,\theta_2,\phi_2)\end{aligned}$$

$$E^{(0)}=E_1+E_2=-\frac{Z^2}{2}\left(\frac{1}{n_1^2}+\frac{1}{n_2^2}\right) \qquad \begin{aligned}n_1&=1,2,3,...\\ n_2&=1,2,3,...\end{aligned} \tag{8-25}$$

The ground-state for the unperturbed system corresponds to when both electrons are in the 1s state. The unperturbed wavefunction in Equation 8-24 becomes the product of two hydrogen 1s wavefunctions.

$$\psi^{(0)}_{1s^2}=\frac{Z^{\frac{3}{2}}}{\sqrt{\pi}}e^{-Zr_1}\cdot\frac{Z^{\frac{3}{2}}}{\sqrt{\pi}}e^{-Zr_2} \tag{8-26}$$

The superscript on the 1s in Equation 8-26 indicates that both electrons are in the 1s state. The unperturbed ground-state energy for a helium atom (Z = 2) is obtained from Equation 8-25.

$$E^{(0)}_{1s^2}=-Z^2=-4hartrees=-108.8eV$$

The unperturbed ground-state energy of helium is now compared to its true energy. This will determine how large of an affect the perturbation of electron-electron repulsion has on the eigenvalues of a two-electron system. The experimental value for the first ionization potential for helium is 24.6 eV. The second ionization potential can be calculated directly from the energy of a one-electron system in Equation 8-13 since this corresponds to a He^+ ion. The second ionization potential is equal to 2 hartrees or 54.4 eV. The true ground-state energy of a helium atom is -24.6 eV + (-54.4 eV) = -79 eV. The unperturbed ground-state energy of a helium atom is in error by approximately 38%. As expected, the electron-electron repulsion term $1/r_{12}$ has a significant affect on the energy of a two-electron system.

The first-order correction to the ground-state energy for a two-electron system can be calculated using Equation 4-13.

$$E_{1s^2}^{(1)} = \int \psi_{1s^2}^{(0)*} \hat{H}^{(1)} \psi_{1s^2}^{(0)} d\tau_1 d\tau_2 \tag{8-27}$$

The volume elements $d\tau_1$ and $d\tau_2$ are the volume elements for each electron. Equation 8-27 can be expanded into the following expression.

$$E_{1s^2}^{(1)} = \frac{Z^6}{\pi^2} \int_0^{2\pi}\int_0^{2\pi}\int_0^{\pi}\int_0^{\pi}\int_0^{\infty}\int_0^{\infty} e^{-2Zr_1} e^{-2Zr_2} \frac{1}{r_{12}} r_1^2 \sin\theta_1 r_2^2 \sin\theta_2 dr_1 dr_2 d\theta_1 d\theta_2 d\phi_1 d\phi_2 \tag{8-28}$$

In order to solve Equation 8-28, an expression for $1/r_{12}$ is needed in terms of the coordinates for each electron.

A convenient way of expressing $1/r_{12}$ is in terms of the spherical harmonics. The details of this expansion can be found in H. Eyring, J. Walter, and G. E. Kimball, *Quantum Chemistry*, Wiley, New York, 1944.

$$\frac{1}{r_{12}} = \begin{cases} \dfrac{1}{r_2} \sum\limits_{l=0}^{\infty} \sum\limits_{m=-l}^{l} \dfrac{4\pi}{2l+1} \left(\dfrac{r_1}{r_2}\right)^l [Y_l^{m*}(\theta_1,\phi_1)][Y_l^m(\theta_1,\phi_2)] & r_2 \geq r_1 \\[2ex] \dfrac{1}{r_1} \sum\limits_{l=0}^{\infty} \sum\limits_{m=-l}^{l} \dfrac{4\pi}{2l+1} \left(\dfrac{r_2}{r_1}\right)^l [Y_l^{m*}(\theta_1,\phi_1)][Y_l^m(\theta_1,\phi_2)] & r_1 \geq r_2 \end{cases} \tag{8-29}$$

The orthonormality of the spherical harmonics will result in a significant amount of cancellations in the expansion of $1/r_{12}$.

Equation 8-29 is now substituted into Equation 8-28. This results in the following integrals to be solved where $r_<$ corresponds to the smaller of r_1 and r_2 and $r_>$ corresponds to the larger of r_1 and r_2.

$$E_{1s^2}^{(1)} = \frac{Z^6}{\pi^2} \sum_{l=0}^{\infty} \sum_{m=-l}^{l} \frac{1}{2l+1} \int_0^{\infty}\int_0^{\infty} e^{-2Zr_1} e^{-2Zr_2} \frac{r_<^l}{r_>^{l+1}} r_1^2 r_2^2 dr_1 dr_2$$

$$x \int_0^{2\pi}\int_0^{\pi} [Y_l^{m*}(\theta_1,\phi_1)][Y_0^0(\theta_1,\phi_1)] \sin\theta_1 d\theta_1 d\phi_1 \quad (8\text{-}30)$$

$$x \int_0^{2\pi}\int_0^{\pi} [Y_0^{0*}(\theta_2,\phi_2)][Y_l^m(\theta_2,\phi_2)] \sin\theta_2 d\theta_2 d\phi_2$$

Due to the orthonormality of the spherical harmonic wavefunctions, the integral in Equation 8-30 is significantly reduced. Taking the limits on r_1 and r_2 for when they are larger than one another, the first-order energy correction is determined by the following expression.

$$E_{1s^2}^{(1)} = \frac{Z^6}{\pi^2} \int_0^{\infty} \left[r_2 e^{-2Zr_2} \left(\int_0^{r_2} r_1^2 e^{-2Zr_1} dr_1 \right) \right] dr_2$$

$$+ \frac{Z^6}{\pi^2} \int_0^{\infty} \left[r_2^2 e^{-2Zr_2} \left(\int_0^{\infty} r_1 e^{-2Zr_1} dr_1 \right) \right] dr_2 \quad (8\text{-}31)$$

Solution of Equation 8-31 results in the following first-order energy correction for the ground-state of a two-electron system.

$$E_{1s^2}^{(1)} (hartrees) = +\frac{5Z}{8} \quad (8\text{-}32)$$

The first-order energy correction for a helium atom is equal to +1.25 hartrees or +34.0 eV. The energy of the ground-state of a helium atom to a first-order correction is now

$$E_{1s^2} = E_{1s^2}^{(0)} + E_{1s^2}^{(1)} = -108.8eV + 34eV = -74.8eV .$$

As mentioned previously, the experimental value for the ground-state energy of a helium atom is -79.0 eV making the computed value to a first-order correction in error by 5.3%. Obtaining the first-order correction to the ground-state energy has significantly improved the computational result; however, the error is still unacceptable.

To improve the computational result, a second-order energy correction is needed. As can be seen by Equation 4-19, obtaining the second-order energy correction to the ground-state will involve mixing higher states. The same is true for obtaining the first-order correction to the wavefunction (see Equation 4-17). The mixing of other configurations into the unperturbed wavefunction is called ***configuration interaction***. The subsequent second-order energy correction for the ground-state of a helium atom is -4.3 eV. A third-order energy correction for the ground-state of a helium atom further improves the result yielding +0.1 eV. The sum of the zero, first, second, and third order energy for the ground-state of a helium atom is -79.0 eV in excellent agreement with the experimental value.

When perturbation theory is applied to obtaining the energy of excited states of a two-electron system, the electron-electron repulsion results in a loss of degeneracy of higher states. The 2p orbitals are no longer degenerate with the 2s orbitals, the 3d orbitals are no longer degenerate with the 3p or 3s orbitals, and so on. The loss of degeneracy as a result of a perturbation is not uncommon.

Another approach to solving the two-electron system is to use variation theory. Part of the affect of having two (or more) electrons in the system is that the electrons tend to shield the nuclear charge from each other. Variation theory can be used to determine the amount of shielding by using two hydrogen wavefunctions (1s) with an adjustable parameter as an ***effective nuclear charge***, ζ.

$$\psi_{trial} = \varphi_1 \varphi_2 = \frac{\varsigma^3}{\pi} e^{-\varsigma r_1} e^{-\varsigma r_2} \tag{8-33}$$

Since φ_1 and φ_2 are orthonormal, ψ_{trial} is also orthonormal. For the case of a helium atom, the value of ζ will presumably be less than two.

The Hamiltonian for the two-electron system can be arranged in a similar fashion as to what was done for the perturbation theory approach in Equations 8-20 through 8-23. The term Z remains the atomic number.

$$\hat{H}\varphi_1\varphi_2 = [\hat{H}_1\varphi_1]\varphi_2 + [\hat{H}_2\varphi_2]\varphi_1 + \frac{\varphi_1\varphi_2}{r_{12}} = E\varphi_1\varphi_2 \tag{8-34}$$

The φ_1 and φ_2 can now be applied to $\hat{H}_1$ and $\hat{H}_2$.

$$\hat{H}_1\varphi_1 = -\frac{1}{2}\nabla_1^2\varphi_1 - \frac{Z}{r_1}\varphi_1 = \frac{(\varsigma - Z)}{r_1}\varphi_1 - \frac{\varsigma^2}{2}\varphi_1 \qquad (8\text{-}35)$$

$$\hat{H}_2\varphi_2 = -\frac{1}{2}\nabla_2^2\varphi_2 - \frac{Z}{r_2}\varphi_2 = \frac{(\varsigma - Z)}{r_2}\varphi_2 - \frac{\varsigma^2}{2}\varphi_2 \qquad (8\text{-}36)$$

Substituting Equations 8-35 and 8-36 into Equation 8-34 results in the following expression.

$$\frac{(\varsigma - Z)}{r_1}\varphi_1\varphi_2 - \frac{\varsigma^2}{2}\varphi_1\varphi_2 + \frac{(\varsigma - Z)}{r_2}\varphi_1\varphi_2 - \frac{\varsigma^2}{2}\varphi_1\varphi_2 + \frac{1}{r_{12}}\varphi_1\varphi_2 = E\varphi_1\varphi_2$$

$$\frac{(\varsigma - Z)}{r_1}\varphi_1\varphi_2 + \frac{(\varsigma - Z)}{r_2}\varphi_1\varphi_2 - \varsigma^2\varphi_1\varphi_2 + \frac{1}{r_{12}}\varphi_1\varphi_2 = E\varphi_1\varphi_2 \qquad (8\text{-}37)$$

Equation 8-37 is now multiplied by $\varphi_1^*\varphi_2^*$ and integrated over all space. This results in the following expression due to the orthonormality of φ_1 and φ_2.

$$E = (\varsigma - Z)\int \varphi_1^*\varphi_2^* \frac{1}{r_1}\varphi_1\varphi_2 d\tau_1 + (\varsigma - Z)\int \varphi_1^*\varphi_2^* \frac{1}{r_2}\varphi_1\varphi_2 d\tau_2$$
$$- \int \varphi_1^*\varphi_2^* \frac{1}{r_{12}}\varphi_1\varphi_2 d\tau - \varsigma^2 \qquad (8\text{-}38)$$

The first two integrals in Equation 8-38 correspond to the average Coulombic potential energy of each electron with the nucleus.

$$(\varsigma - Z)\int \varphi_1^*\varphi_2^* \frac{1}{r_1}\varphi_1\varphi_2 d\tau_1 = (\varsigma - Z)\int \varphi_1^*\varphi_2^* \frac{1}{r_2}\varphi_1\varphi_2 d\tau_2 = \varsigma^2 - Z\varsigma \qquad (8\text{-}39)$$

The third integral in Equation 8-38 is the electron-electron repulsion potential. This is the same as the integral that was solved previously in the perturbation theory method (see Equations 8-27 through 8-32) except now Z is replaced with ζ.

$$\int \varphi_1^* \varphi_2^* \frac{1}{r_{12}} \varphi_1 \varphi_2 d\tau = \frac{5\varsigma}{8} \tag{8-40}$$

Equations 8-39 and 8-40 can now be substituted into Equation 8-38 resulting in an energy expression in terms of the adjustable parameter ζ.

$$E = \varsigma^2 - 2Z\varsigma + \frac{5\varsigma}{8} \tag{8-41}$$

Now that an energy expression has been obtained in terms of the effective nuclear charge ζ, an optimal value for ζ must be determined by minimizing the energy.

$$\frac{dE}{d\varsigma} = 0 = 2\varsigma - 2Z + \frac{5}{8}$$

$$\varsigma = Z - \frac{5}{16} \tag{8-42}$$

For a helium atom Z = 2, so the effective nuclear charge is equal to 27/16. Physically this means that the electrons experience a nuclear charge of 27/16 rather than 2. If the optimized value of ζ for helium is substituted into the Equation 8-41, the energy eigenvalue for the ground-state of helium is -2.85 hartrees = -77.5 eV which is in relatively good agreement with experiment.

The perturbation and variational approach to solving for the energy of a helium atom demonstrates that the hydrogen atom wavefunctions are not a good starting point for solving the Schroedinger equation of atoms with multiple electrons. The electron-electron repulsion potential has a profound affect on the energy of a system with multiple electrons, as it has been determined for the case of the helium atom. The charge of the nucleus experienced by the electrons is reduced as a result of shielding, and some of the degeneracy of the orbitals is lost. A better set of functions as a basis set for solving systems with multiple electrons will be discussed in Section 8.4.

8.3 ELECTRON SPIN

The electrons in an atom also have an *intrinsic* angular momentum in addition to their orbital angular momentum about the nucleus. This is called the ***electron spin*** or sometimes just referred to as ***spin***. Even an electron in the $l = 0$ orbital that has zero angular momentum will have an intrinsic spin. The intrinsic spin of the electron is not a classical mechanical effect; hence, it is not a correct picture to view the electron spinning about one of its axes, as the classical mechanical picture would indicate. The term "spin" is more of a name for this phenomenon rather than an actual description of the electron. Though the intrinsic spin of the electron is real, there is no example in the macroscopic world to form a visual model. The electron spin arises naturally when relativistic mechanics is combined with quantum mechanics. Since this text is confined to quantum mechanics, the concept of electron spin must be introduced as a hypothesis.

Since an electron has an intrinsic spin, there must be a corresponding operator for the overall intrinsic spin angular momentum squared, $\hat{S}^2$. It is expected that the intrinsic spin eigenfunctions, λ_{SM}, are analogous to the spatial spherical harmonic wavefunctions, $Y_{lm}(\theta, \phi)$. The operators $\hat{S}_z$ and $\hat{S}^2$ will be the only operators for which the intrinsic spin functions are eigenfunctions just like $Y_{lm}(\theta, \phi)$ are only eigenfunctions of $\hat{L}^2$ and $\hat{L}_z$ operators.

$$\hat{S}^2 \lambda_{SM} = S(S+1)\hbar^2 \lambda_{SM} \tag{8-43}$$

$$\hat{S}_z \lambda_{SM} = \hbar M \lambda_{SM} \tag{8-44}$$

Equation 8-43 is the analog of the equation for overall orbital angular momentum squared.

$$\hat{L}^2 Y_{lm} = l(l+1)\hbar^2 Y_{lm}$$

Equation 8-44 is the analog to the equation for the z-component of orbital angular momentum.

$$\hat{L}_z Y_{lm} = m\hbar Y_{lm}$$

There are only two possible values for the M and S quantum numbers for a single electron (such as in a hydrogen atom): +½ or -½. The eigenfunction for the M = +½ eigenstate is given the symbol α and is called "spin up". The M = -½ eigenstate is given the symbol β and is called "spin down".

$$\hat{S}^2\alpha = \hbar^2\left(\frac{3}{4}\right)\alpha \qquad \hat{S}_z\alpha = \left(\frac{\hbar}{2}\right)\alpha \tag{8-45}$$

$$\hat{S}^2\beta = \hbar^2\left(\frac{3}{4}\right)\beta \qquad \hat{S}_z\beta = -\left(\frac{\hbar}{2}\right)\beta \tag{8-46}$$

The complete designation for a hydrogen atom wavefunction will include the intrinsic spin eigenstate: $\psi_{n,l,m,\alpha}$ or $\psi_{n,l,m,\beta}$.

The ***Pauli principle*** states that no two electrons in an atom or molecule can occupy the same spin-orbital. This means that for an atom, each spatial orbital (e.g. 1s, 2s, $2p_0$, $2p_1$, $2p_{-1}$, and so on) can have only two electrons and they must be of opposite spin. This adds a two-fold degeneracy to each spatial orbital for an atom or a molecule.

8.4 COMPLEX ATOMS

The Hamiltonian for an atom with N electrons, ignoring nuclear motion, can be written as follows.

$$\hat{H} = -\frac{1}{2}\sum_{i=1}^{N}\nabla_i^2 - \sum_{i=1}^{N}\frac{Z}{r_i} + \sum_{i=1}^{N}\sum_{j>i}\frac{1}{r_{ij}} \tag{8-47}$$

The first term in the Hamiltonian corresponds to the kinetic energy of each electron, the second term is the Coulombic attraction of each electron to the nucleus with an atomic number Z, and the third term is the Coulombic repulsion between each electron. The index $j > i$ in the summation prevents terms such as $1/r_{ii}$.

The zeroth-order wavefunction, as in the case for the helium atom, will be a product of N-one-electron functions.

$$\psi^{(0)} = f_1(r_1,\theta_1,\phi_1)f_2(r_2,\theta_2,\phi_2)\ldots f_N(r_N,\theta_N,\phi_N) \tag{8-48}$$

The general form of the functions f will be a radial function, $R_{nl}(r)$, times a spherical harmonic function, $Y_{lm}(\theta, \phi)$.

$$f = R_{nl}(r)Y_{lm}(\theta,\phi) \tag{8-49}$$

One possibility as a basis set of functions to be used for f is the hydrogen-atom functions. As seen in the case of a helium atom, this is not a particularly good start. The hydrogen-atom wavefunctions do not account for shielding and other affects of the inter-electronic repulsion. A basis set of functions that take this into account is a much better starting point for the calculation. J. C. Slater created such a basis set of functions known as the ***Slater-type orbitals (STO)***. The functions have the following general form.

$$\psi_{nlm} = Nr^{n^*-1}e^{-(Z-s)r/n^*}Y_{lm}(\theta,\phi) \tag{8-50}$$

The term s is the shielding constant and n^* is a parameter that varies with the principal quantum number n. The term N is the radial normalization constant. The effective nuclear charge, ς, can be calculated from s and n^* as follows.

$$\varsigma = \frac{Z-s}{n^*} \tag{8-51}$$

The Slater-type orbitals replace the polynomial in r as in hydrogenlike orbitals with a single power in r reducing computational effort. The values for s and n^* are determined empirically by the following procedure.

1. The electrons of the atom are put into the following groups.

   {1s}; {2s, 2p}; {3s, 3p}; {3d}; {4s, 4p}; {4d}; {4f}; {5s, 5p}; ...

2. There is no contribution to screening, s, from any electron within a given group.

3. In the 1s group, the contribution to s is 0.30. For electrons outside the 1s group, the contribution to s is 0.35 for each electron in that group.
4. For an electron in an s or p orbital, the contribution to s is 0.85 for each other electron when the principal quantum number n is one less than for the orbital being written. For still lower levels of n, the contribution to s is 1.00.
5. For electrons in d and f orbitals, the contribution to s is 1.00 for each electron below the one for which the wavefunction is being written.
6. The value for n^* is determined based on the value for n as follows.

n =	1	2	3	4	5	6
n^* =	1	2	3	3.7	4.0	4.2

Example 8-4

Problem: Determine the Slater-type orbital wavefunction and for an electron in a) the ground-state of helium, and b) the $2p_m$ orbital of oxygen.

Solution:

a) For helium, Z = 2. The only screening is from the other electron so value for s = 0.3. The value of n = 1, so the value for n* = 1. The Slater-type orbital wavefunction for a helium atom in the ground-state is as follows.

$$\psi_{100} = Ne^{-1.7r}Y_{00}(\theta,\phi)$$

The effective nuclear charge is 1.7, the same value as obtained previously from variational theory in Section 8.2.

b) For oxygen, Z = 8. For an electron in the $2p_m$ orbital, n = 2 and so n* = 2. The contributions to the screening constant s are summed as follows.

2 electrons in the 1s orbital:	2(0.85) = 1.7
5 electrons in 2s and 2p orbitals:	5(0.35) = 1.75

The total for s is 3.45. The wavefunction for the $2p_m$ orbital in oxygen has the following form.

$$\psi_{2,1,m} = re^{-2.28r}Y_{1m}(\theta,\phi)$$

The effective nuclear charge for an electron in a $2p_m$ orbital in oxygen is 2.28.

There are several deficiencies in STO's. Because STO's replace the polynomial in r for a single term, STO's do not have the proper number of nodes and do not represent the inner part of an orbital well. Care must be taken when using STO's because orbitals with the different values of n but the same values of l and m_l are not orthogonal to one another. Another deficiency is that ns orbitals where $n > 1$ have zero amplitude at the nucleus. Values have been obtained for the effective nuclear charge for a number of atoms by fitting STO's to numerically computed wavefunctions. These values are given Table 8-2 and supersede the values obtained empirically from Slater's rules.

Table 8-2. The numerical values for the effective nuclear charge for atoms in a number of neutral ground-state atoms are shown. Values were obtained from E. Clementi, D. L. Raimondi, IBM Research Note NJ-27, 1963.

	H							**He**
1s	1							1.6875
	Li	**Be**	**B**	**C**	**N**	**O**	**F**	**Ne**
1s	2.6906	3.6848	4.6795	5.6727	6.6651	7.6579	8.6501	9.6421
2s	1.2792	1.9120	2.5762	3.2166	3.874	4.4916	5.1276	5.7584
2p			2.4214	3.1358	3.8340	4.4532	5.1000	5.7584
	Na	**Mg**	**Al**	**Si**	**P**	**S**	**Cl**	**Ar**
1s	10.6529	11.6089	12.5910	13.5754	14.5578	15.5409	16.5239	17.5075
2s	6.5714	7.3920	8.2136	9.0200	9.8250	10.6288	11.4304	12.2304
2p	6.8018	7.8258	8.9634	9.9450	10.9612	11.9770	12.9932	14.0082
3s	2.5074	3.3075	4.1172	4.9032	5.6418	6.3669	7.0683	7.7568
3p			4.0656	4.2852	4.8864	5.4819	6.1161	6.7641

Solving the Schroedinger equation for an atom with N electrons is a formidable computational task because of the numerous electron-electron repulsion terms, $1/r_{ij}$. In order to calculate the electron repulsion of one electron, the wavefunctions for the other electrons must be known and vice versa. The best atomic orbitals are obtained by a numerical solution of the Schroedinger equation. The procedure first introduced by D.R. Hartree is called ***self-consistent field (SCF)***. The procedure was further improved by including electron exchange by V. Fock and J.C. Slater. The orbitals obtained by a combination of these procedures are called ***Hartree-Fock self-consistent field orbitals***.

The ***Hartree-Fock self-consistent field*** (HF-SCF) approach assumes that any one electron moves in a potential that is a spherical average due to the other electrons and the nucleus. The spherically averaged potential for an electron is expressed as a single charge that is centered on the nucleus and varies with the position r in the potentially averaged sphere. The Schroedinger equation is then numerically solved for that electron in the spherically averaged potential. Of course in order to determine the spherically averaged potential for a particular electron, the wavefunctions (and hence relative positions) of the other electrons must be known. Since the wavefunctions of the other atoms is most likely not known, the calculations begin with approximate wavefunctions as a basis set for the other electrons such as STO's. The wavefunction is assumed to be a product of one-electron wavefunctions as in Equation 8-48. The result of this assumption is that the electrons in the atom are ordered in hydrogenlike orbitals. As an example, the electrons in oxygen (Z = 8) are ordered in the familiar fashion of $1s^2 2s^2 2p^4$. The Schroedinger equation is then solved for the electron, and then the procedure is repeated for the rest of the electrons in the atom. After this first computation, a set of improved wavefunctions as the basis set for the electrons is obtained. The computation is now repeated with this new set of wavefunctions for each electron. A new set of wavefunctions is obtained for each electron and is compared to wavefunctions from the previous computational cycle. If the values are different, a new computational cycle is performed with the latest wavefunctions obtained for the electrons. If the wavefunctions do not differ significantly from the previous computational cycle, the computation is complete and the wavefunctions are said to be self-consistent.

The details of a HF-SCF computation can now be examined in detail. In the HF-SCF approach, the Hamiltonian for an atom is written in terms of a summation of hydrogenlike terms plus the electron repulsion terms.

$$\hat{H} = \sum_{i}^{N} \hat{H}_i^{core} + \frac{1}{2}\sum_{ij}{}' \frac{1}{r_{ij}} \tag{8-52}$$

The term $\hat{H}_i^{core}$ is called the ***core Hamiltonian*** and represents the electron i in a potential that consists of only the nucleus of atomic number Z with no repulsive potential from any other electron (as in a one-electron system). The factor of ½ is to eliminate counting the same electron-electron repulsions twice. The prime is a reminder not to count any $1/r_{ii}$ terms.

The focus is now on electron 1, and the rest of the electrons (2, 3, 4, ..., N) are regarded as being distributed about to form part of the spherically averaged potential that electron 1 travels through. The charge of a given electron is smeared out into a continuous charge density, ρ_i, (the charge of an electron per unit volume) that electron 1 travels through. The potential of electron 1 with another electron, V_{1i}, is obtained by summing the product of the charge of electron 1 times an infinitesimal charge density dρ_i times an infinitesimal volume element, dv_i, divided by the distance of separation, r_{1i}.

$$V_{1i} = \int \frac{\rho_i}{r_{1i}} dv_i \tag{8-53}$$

The probability density of the electrons is given as $|s_i|^2$. As a result, the charge density of an electron is given as $\rho_i = -|s_i|^2$.

$$V_{1i} = \int \frac{|s_i|^2}{r_{1i}} dv_i \tag{8-54}$$

The potential interaction of electron 1 with all N electrons are determined and summed together.

$$V(r_1, \theta_1, \phi_1) = V_{12} + V_{13} + V_{14} + \cdots + V_{1N} = \sum_{i=2}^{N} \int \frac{|s_i|^2}{r_{1i}} dv_i \tag{8-55}$$

At this point, the HF-SCF approach makes an additional assumption beyond assuming that the wavefunction is a product of one-electron wavefunctions. It is assumed that the potential of an electron in an atom can be made into a function of *r* only. This is called the ***central-field approximation***. The potential $V(r_1, \theta_1, \phi_1)$ is averaged over θ and ϕ so that it is a function of r_1 only.

$$V(r_1) = \frac{\int_0^{2\pi}\int_0^{\pi} V(r_1, \theta_1, \phi_1) \sin\theta_1 d\theta_1 d\phi_1}{\int_0^{2\pi}\int_0^{\pi} \sin\theta d\theta d\phi} \tag{8-56}$$

The one-electron Schroedinger equation for electron 1 is now solved and an improved wavefunction χ_1 for electron 1 is obtained.

$$[\hat{H}_1^{core} + V(r_1)]\chi_1(1) = \varepsilon_1 \chi_1(1) \tag{8-57}$$

The energy eigenvalue ε_1 is energy of electron 1 at this stage of the approximation. The procedure is continued for all N electrons in the atom. The wavefunctions for each electron, χ_i, are compared to the original wavefunctions at the beginning of the calculation. If the χ_i wavefunctions do differ significantly, the computation is complete and SCF has been achieved. If not, another computational cycle is performed.

Once the final self-consistent field wavefunctions are obtained, the Hartree-Fock energy can now be obtained. A tempting conclusion at this point is to simply take a sum of the energy eigenvalues obtained for each electron: $E = \varepsilon_1 + \varepsilon_2 + \varepsilon_3 + \ldots + \varepsilon_N$. This is would be an incorrect assumption because the energy eigenvalues were obtained by first determining the potential average between each electron. In calculating ε_1, it is determined by getting the electron-electron repulsion between electrons 1 and 2, 1 and 3, all the way to 1 and N. In calculating ε_2, the electron repulsion is determined between electrons 2 and 1, 2 and 3, all the way to 2 and N. As can be seen, the electron-electron repulsions are over counted if the energy is determined merely by the sum of ε_i's. The repetitive electron-electron repulsions must be subtracted from the sum

$$E = \sum_{i=1}^{N} \varepsilon_i - \sum_{i=1}^{N} \sum_{j>i} \int\int \frac{|\psi_i(i)|^2 |\psi_j(j)|^2}{r_{ij}} dv_i dv_j \quad (8\text{-}58)$$

The HF-SCF atomic orbitals are not the best that can be obtained. The approximation is rooted in the orbital picture for the individual electrons and in central-field approximation for the potential. The electron densities calculated from HF-SCF are quite accurate, but the energy eigenvalues are too high. As an example, the HF-SCF ground-state energy for a helium atom is -77.9 eV compared to the experimental value of -79.0 eV. In order to improve the calculated result, the separation of the electron motions approximation must be relaxed and r_{ij} must be incorporated into the wavefunctions. This is called the ***correlation problem***, and this is discussed further in the next chapter. For heavier elements, relativistic effects also need to be included into the Schroedinger equation. Relativistic effects are important in describing certain properties of heavier elements such as the color of gold, the liquid form of mercury, and the contraction of lanthanide.

8.5 SPIN-ORBIT INTERACTION

The electrons in an atom contain angular momentum (except when $l = 0$) and an intrinsic spin. According to classical electromagnetic theory, when a charge q moves in a circular path, a magnetic field is generated that is associated with the magnetic dipole source. The magnetic dipole moment, $\vec{\mu}$, from the charge flowing through a circular loop is proportional to the current and the area of the loop. The direction of the magnetic dipole moment is perpendicular to the plane of the loop. An electron in an orbit around the nucleus can be considered as a negative charge of -e flowing around a loop of some radius r generalizing the true orbital motion. The area of the loop is πr^2. The current is the frequency, ω, that the electron passes through a particular point in the loop ($\omega/2\pi$).

$$\mu = \pi r^2 \left(\frac{-e\omega}{2\pi} \right) \frac{1}{c} = -\frac{er^2\omega}{2c} \quad (8\text{-}59)$$

The term c in Equation 8-59 is the speed of light. The angular momentum of a particle moving about a circular loop is the particle's mass times the square of the radius of the loop times the frequency, $mr^2\omega$. The angular momentum of an electron in an orbit is determined by the $\hat{L}^2$ and $\hat{L}_z$ operators; hence, Equation 8-59 can be rewritten as follows for an electron in a hydrogen orbital.

$$\vec{\mu} = -\frac{e\vec{L}}{2m_e c} = -\frac{e}{2m_e c}(\hat{L}_x + \hat{L}_y + \hat{L}_z) \tag{8-60}$$

As can be seen by Equation 8-60, the magnetic dipole moment is proportional to the angular momentum of the electron. Since the angular momentum of an electron will be in units of $\hbar$, it is convenient to collect the constant terms in Equation 8-60 and define a new constant called the ***Bohr magneton***, μ_B.

$$\mu_B \equiv \frac{e\hbar}{2m_e c} \tag{8-61}$$

Electronic magnetic dipole moments in molecules and atoms are measured in terms of Bohr magnetons in the same way that angular momentum is measured in terms of $\hbar$.

The same analysis can now be done for the intrinsic spin of an electron. The magnetic moment as a result of the intrinsic spin will be directly proportional to the angular momentum of the intrinsic spin, $\vec{S}$.

$$\vec{\mu} = g_e \frac{e}{2m_e c}\vec{S} \tag{8-62}$$

The expression for the magnetic dipole moment for the intrinsic spin of an electron is similar to that of an electron in its orbit except that an additional term g_e is needed. The additional term is needed because the simple model of a circulating electron used to obtain Equation 8-60 does not apply to the intrinsic spin of an electron.

When an external magnetic field is applied to an atom, the effect of the field must be incorporated into the Schroedinger equation. In classical echanics, the interaction of the magnetic dipole moment and the external agnetic field, $\vec{H}$, is determined by the dot product. Using the classical echanical description, the effect of an external magnetic field on a ydrogen atom is obtained by taking the dot product of $\vec{H}$ and the xpression for $\vec{\mu}$ in Equation 8-60.

$$\vec{\mu} \cdot \vec{H} = -\frac{e}{2m_e c}(\hat{L}_x H_x + \hat{L}_y H_y + \hat{L}_z H_z)$$

f the external magnetic field is applied uniformly only along the z-axis (H_z s a constant), the terms along the x and y-axes are zero. The additional term dded to the Hamiltonian for the hydrogen becomes as follows.

$$\hat{H}^{magnetic} = -\vec{\mu} \cdot \vec{H} = \frac{e}{2m_e c} H_z \hat{L}_z = \frac{\mu_B H_z}{\hbar} \hat{L}_z \qquad (8\text{-}63)$$

he Hamiltonian for a hydrogen atom in a uniform magnetic field along the -axis can be written as follows where $\hat{H}^{(0)}$ is the unperturbed hydrogen tom Hamiltonian.

$$\hat{H} = \hat{H}^{(0)} + \hat{H}^{magnetic} = \hat{H}^{(0)} + \frac{\mu_B H_z}{\hbar} \hat{L}_z \qquad (8\text{-}64)$$

The wavefunctions for a hydrogen atom, ψ_{nlm}, are eigenfunctions of $\hat{H}^{(0)}$ nd $\hat{L}_z$. As a result, ψ_{nlm} is an eigenfunction of $\hat{H}$, the Hamiltonian of a ydrogen atom in a magnetic field.

$$\left(\hat{H}^{(0)} + \frac{\mu_B H_z}{\hbar} \hat{L}_z\right)\psi_{nlm} = (E_n + m\mu_B H_z)\psi_{nlm} \qquad (8\text{-}65)$$

he energy of a hydrogen atom in an applied magnetic field depends on the m_l quantum number. The magnetic field removes the degeneracy of states vith the same n and l but with different m_l quantum numbers. The removing f degenerate levels as a result of an external magnetic field is called the ***eeman effect.*** According to Equation 8-65, the separation between the ifferent m_l levels will increase with increasing strength of the magnetic field, H_z. For this reason, the m_l quantum number is referred to as the ***magnetic quantum number***.

Both the orbital motion and the intrinsic spin of electron have a magnetic dipole associated with it. The two magnetic dipoles may interact with one another. This feature of atomic and molecular structure is called ***spin-orbit interaction***. The spin-orbit interaction is a coupling of the two different "motions" of spin of the electron. The Hamiltonian that describes this interaction is a dot product of the angular momentum vectors of the two different types of spin "motions" of the electron. The proportionality constant is α which can be measured by spectroscopy.

$$\hat{H}^{spin-orbit} = \alpha\vec{L} \cdot \vec{S} \qquad (8\text{-}66)$$

Equation 8-66 can be rewritten in terms of $\hat{L}^2$ and $\hat{S}^2$ in the following fashion where the total of orbital and spin angular momentum is given as $\vec{J} = \vec{L} + \vec{S}$.

$$\hat{H}^{spin-orbit} = \alpha\left(\vec{J} \cdot \vec{J} + \vec{L} \cdot \vec{L} + \vec{S} \cdot \vec{S}\right)/2 = \frac{\alpha}{2}\left(\hat{J}^2 + \hat{L}^2 + \hat{S}^2\right) \qquad (8\text{-}67)$$

The spin-orbit coupled states will now have an additional quantum number J that refers to the total of orbital and spin angular momentum. The designation of a spin-coupled state is ψ_{nJls} and is an eigenfunction of spin-orbit Hamiltonian. For a hydrogen atom, the eigenvalues of $\hat{L}^2\ \psi_{nSlm}$ are $-\hbar l(l + 1)$, the eigenvalues of $\hat{S}^2\ \psi_{nSlm}$ are -ħS(S + 1) (see Equation 8-44), and the eigenvalues of $\hat{J}^2\ \psi_{nSlm}$ are analogously ħJ(J + 1).

$$\hat{H}^{spin-orbit}\psi_{nJls} = \frac{\alpha\hbar^2}{2}[J(J+1) - l(l+1) - S(S+1)]\psi_{nJls} \qquad (8\text{-}68)$$

The allowed values of J range from $l + S$ downward in steps of one to $|l - S|$. For the case of a hydrogen atom with only one electron, S = ½. The possible J values are equal to $l \pm$ ½. In the case of the ground-state of hydrogen, $l = 0$ and the only possible value of J is J = ½.

The energy of the spin-orbit coupled eigenstates for a hydrogen atom is as follows.

$$E_{nJ,ls} = E_n + \frac{\alpha\hbar^2}{2}[J(J+1) - l(l+1) - S(S+1)] \qquad (8\text{-}69)$$

The result of the coupling of spin and orbit angular momentum is to create an energy difference between states that would otherwise be degenerate. This phenomenon is called ***spin-orbit splitting***. As an example, consider the n = 2, l = 0 state of a hydrogen atom. In the absence of spin-orbit effects, there are six degenerate states: $2p_1\alpha$, $2p_1\beta$, $2p_0\alpha$, $2p_0\beta$, $2p_{-1}\alpha$, and $2p_{-1}\beta$. Due to spin-orbit coupling, these states may be mixed in some way, and the states would identified by the two possible J values: J = 1 + 1/2 = 3/2 and J = 1 - 1/2 = 1/2. Note that are still six states because the J = 3/2 state is four-fold degenerate and the J = 1/2 state is two-fold degenerate. The difference in energy between two spin-orbit coupled states can be determined by using Equation 8-68.

For J = 3/2:

$$E_{2,3/2,1,m} = E_2 + \frac{\alpha\hbar^2}{2}\left\{\frac{3}{2}\left(\frac{3}{2}+1\right) - 1(1+1) - \frac{1}{2}\left(\frac{1}{2}+1\right)\right\} = E_2 + \frac{\alpha\hbar^2}{2}$$

For J = 1/2:

$$E_{2,1/2,1,m} = E_2 + \frac{\alpha\hbar^2}{2}\left\{\frac{1}{2}\left(\frac{1}{2}+1\right) - 1(1+1) - \frac{1}{2}\left(\frac{1}{2}+1\right)\right\} = E_2 - \alpha\hbar^2$$

$$\Delta E = E_{2,3/2,1,m} - E_{2,1/2,1,m} = -\frac{\alpha\hbar^2}{2}$$

If the spin-orbit energy difference between these two spin-orbit coupled energy states is observed in the emission or absorption spectra of a hydrogen atom, the energy difference between spectral lines can be used to obtain the value for α.

In the case of atoms with more than one electron, the spin-orbit interaction is observable in the emission or absorption spectra of the atoms, even though the interaction energies are small relative to the transition energies of the spectral lines. In the case of light elements, the strongest coupling magnetic dipoles is between all those associated with orbital motion with all of those associated with intrinsic spin. So for light elements, the coupling must be found between the total angular momentum vector $\vec{L}$ and the total intrinsic spin vector $\vec{S}$ to form the total angular momentum vector $\vec{J}$. In the case of heavier elements, the strongest coupling of dipole moments occurs between the orbital and intrinsic spin of the individual electrons. The total angular momentum of each electron is determined in the same way as was done previously for a hydrogen atom and then summed over all of the electrons to obtain $\vec{J}$. The coupling for light elements will be discussed in detail here.

To aid in counting the possible intrinsic spin states, it is convenient to group the electrons in ***shells*** and ***subshells***. The spin-orbitals with the same n quantum number are referred to as a shell. A set of spin-orbitals with the same n and l quantum numbers are referred to as a subshell. According to the Pauli principle, a subshell of $l = 0$ can have a maximum occupancy of two electrons; a subshell of $l = 1$ can have a maximum occupancy of six electrons; a subshell of $l = 2$ can have a maximum occupancy of ten electrons; and so forth. Electrons in the same subshell are said to be equivalent, and electrons in different subshells are said to nonequivalent.

As a first example, consider a hypothetical excited electronic state of lithium.

$$1s^1 2p^1 3p^1$$

All three electrons are in different shells and, hence, nonequivalent. The orbital angular momentum of each electron is defined as $\vec{l}_1$, $\vec{l}_2$, and $\vec{l}_3$. The first step is to determine the resultant orbital angular momentum of the first two electrons, $\vec{L}_{12}$. This is determined first because the quantum number for the magnitude of a resultant angular momentum vector may take on the values from the sum of the two sources down to the absolute value of their differences. In this example, $l_1 = 0$ for electron 1 and $l_2 = 1$ for electron 2. The magnitude of the vector sum L_{12} is equal to only 1.

$$\vec{L}_{12} = \vec{l}_1 + \vec{l}_2 \qquad\qquad L_{12} = 1$$

The orbital angular momentum vector of the third electron, $\vec{l}_3$, is now added to $\vec{L}_{12}$.

$$\vec{L}_{12} + \vec{l}_3 = \vec{L}_{total} \qquad L_{total} = 1+1,\ldots,|1-1| = 2,1,0$$

This result means that there are three possibilities in the coupling of orbital angular momentum with intrinsic spin angular momentum.

The total intrinsic spin angular momentum is determined in the same fashion. However, in the case of intrinsic spin angular momentum, the quantum number for the spin is always ½ limiting the total number of possibilities. The magnitude of the intrinsic spin vector, S_{12}, for electrons 1 and 2 are determined first.

$$\vec{S}_{12} = \vec{S}_1 + \vec{S}_2 \qquad S_{12} = \frac{1}{2}+\frac{1}{2},\ldots,\left|\frac{1}{2}-\frac{1}{2}\right| = 1,0$$

This result indicates that two nonequivalent electrons may be coupled to the intrinsic spin in two different ways. The third spin is now added to $\vec{S}_{12}$.

$$\vec{S}_{12} + \vec{S}_3 = \vec{S}_{toal} \qquad S_{total} = \frac{1}{2}+1,\ldots,\left|\frac{1}{2}-1\right|;\ and\ \frac{1}{2}+0 = \frac{3}{2},\frac{1}{2},\frac{1}{2}$$

To get the S_{total}, both possible values of S_{12}, must be included resulting in *two* different ways that intrinsic spin can be coupled with the designation of S_{total} = ½.

Multiplicity associated with angular momentum is always two times the value plus one. The multiplicity of orbital angular momentum is $2l + 1$. The total intrinsic spin multiplicity is $2S + 1$, and the resultant multiplicity of the orbital-spin coupling angular momentum is 2J + 1. Spin multiplicities of 1, 2, 3, and 4 are called singlet, doublet, triplet, and quartet respectively. For the current example of the excited electronic state of lithium, the three nonequivalent electrons may be coupled to form a quartet state ($S = 3/2$):

$$2S + 1 = 2\left(\frac{3}{2}\right) + 1 = 4\,;$$

or may be coupled to form two different doublet states ($S = ½$):

$$2S+1=2\left(\frac{1}{2}\right)+1=2\,.$$

The magnetic moment vector $\vec{J}$ that results from the coupling of the total orbital angular momentum vector $\vec{L}_{total}$ and the total intrinsic spin vector $\vec{S}_{total}$ can now be determined.

$$\vec{J}=\vec{L}_{total}+\vec{S}_{total}\,; \qquad J=(L_{total}+S_{total}),\ldots,\left|L_{total}-S_{total}\right|$$

The number of possible J states must be determined for each combination of possible S_{total} and L_{total}. For the case of the excited state of lithium, the following J coupled states are possible.

L_{total}	S_{total}	Possible J Values
0	1/2	1/2
0	1/2	1/2
0	3/2	3/2
1	1/2	3/2,1/2
1	1/2	3/2,1/2
2	1/2	5/2,3/2
2	1/2	5/2,3/2
2	3/2	7/2,5/2,3/2,1/2

As can be seen for the case of three nonequivalent electrons in the excited electronic state of lithium, there are many distinct spin-orbit coupled states possible.

The energy of the various spin-orbit coupled states can be determined in the same fashion as for the hydrogen atom (see Equation 8-69).

$$E_{JLS}^{spin-orbit}=\frac{\alpha\hbar^2}{2}[J(J+1)-L_{total}(L_{total}+1)-S_{total}(S_{total}+1) \qquad (8\text{-}70)$$

The value of α is determined by experiment and will vary for different systems. As can be seen by Equation 8-70, states with a large intrinsic spin multiplicity will be of lower energy. For states with the same intrinsic spin multiplicity, the lowest energy level will be state with largest L_{total}. The energy ordering of spin-orbital coupled states were first observed on the basis of atomic spectra and is called ***Hund's rules***.

The various electronic states are designated using ***term symbols***. The term symbols state the values for J, L, and S. For the value of L, a letter symbol is used: for $L = 0$, the symbol S is used; for $L = 1$, the symbol P is used; for $L = 1$, the symbol P is used; and the symbols G, H, I through the alphabet for subsequent increasing values of L. A superscript on the left side of the L symbol designates the intrinsic spin multiplicity, and a subscript designates the particular spin-orbit coupled J state.

$$^{(2S+1)}L_{\mathrm{J}}$$

As an example, for the lowest spin-orbit coupled state for the electronically excited lithium atom $(1s^1 2p^1 3p^1)$, $L_{total} = 2$, $S_{total} = 3/2$, and $\mathrm{J} = 1/2$, the term symbol is $^4P_{1/2}$. These are called ***Russell-Saunders*** term symbols because it is assumed that the individual orbital angular momentum are more strongly coupled than the spin-orbit coupling. If spin-orbit coupling is ignored, the J term is omitted from the term symbol.

Example 8-5

Problem: Determine the term symbol for a hydrogen atom ignoring spin-orbit coupling in a) ground-state, b) the 2s orbital, c) the 2p orbital, and d) a 3d orbital.

Solution:

a) The ground-state: 1s. For a 1s orbital, $l = 0$ and $S = 1/2$. The term symbol is ^2S.

b) For a 2s orbital, $l = 0$ and $S = 1/2$. The term symbol is again ^2S.

c) For a 2p orbital, $l = 1$ and $S = 1/2$. The term symbol is ^2P.

d) For a 3d orbital, $l = 2$ and $S = 1/2$. The term symbol is ^2D.

In summary, the Hamiltonian of an atom exposed to a magnetic field (Zeeman effect) can be broken down into the following components: the core Hamiltonian, $\hat{H}^{core}$; the electron-electron repulsion, $\hat{H}^{rep}$; the spin-orbit coupling, $\hat{H}^{spin-orbit}$; and the Zeeman effect, $\hat{H}^{magnetic}$

$$\hat{H} = \hat{H}^{core} + \hat{H}^{rep} + \hat{H}^{spin-orbit} + \hat{H}^{magnetic} \tag{8-71}$$

Note that all of the terms in Equation 8-71 are internal properties of the atom except $\hat{H}^{magnetic}$ is due to an externally applied magnetic field. The energy eigenvalues of each state is determined as a sum of these effects that have been developed throughout this chapter. The electron-electron repulsion term has a very large effect on the energy of the system following the spin-orbit coupling and then the Zeeman effect depending on the strength of the applied magnetic field. The contributory affect of each of these properties is shown in Figure 8-5 for the excited electronic state of helium: $1s^1 2p^1$. Note

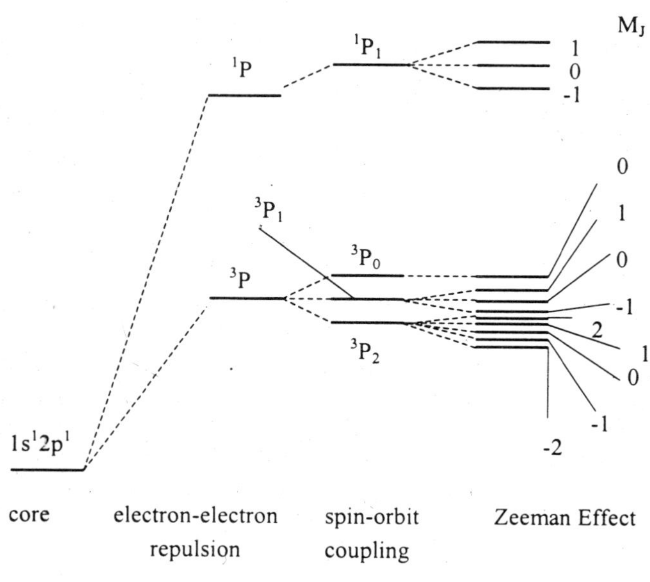

Figure 8-5. The relative individual contributions on the number of possible states for an electronically excited helium atom where the electrons are in the $1s^1 2p^1$ orbitals are shown. All of the effects are internal except the Zeeman effect that is a result of an externally applied magnetic field.

that the number of possible states increases with each contributory term in Equation 8-71.

8.6 SELECTION RULES AND ATOMIC SPECTRA

To cause electronic transitions in atoms from the ground-state will generally require radiation in the visible or ultraviolet region of the electromagnetic spectrum. An atomic spectrum can be in terms of absorption as an electron is promoted from a lower energy orbital to a higher one, or the spectrum can be in terms of emission, following an absorption, as the electron goes from a higher energy orbital to a lower orbital energy.

The selection rules for atomic spectra are determined in the same way as it was done previously for vibration-rotation spectroscopy (see Section 6.7). For a hydrogen atom, the dipole moment operator of the incoming photon, $\hat{H}_t$, is operated on and integrated over all space between the initial state n, l, m_l with the final state n', l', m_l'. Allowed transitions occur when the integral is non-zero and forbidden transtions are when the integral is equal to zero.

$$\langle n, l, m_l | \hat{H}_t | n', l', m_l' \rangle \qquad (8\text{-}72)$$

The selection rules in terms of l and m_l have been determined previously for the vibration-rotation of a molecule (recall that the wavefunction for a hydrogen atom is in part the spherical harmonics, $R_{nl}Y_{lm}$) by simply replacing the J for an l and M_J for an m_l in Equation 6-84.

$$|\Delta l| = 1$$

$$|\Delta m_l| = 0$$

The integral in Equation 8-72 in terms of the radial component is considerably more difficult to solve. The result of the integration is that any Δn is allowed.

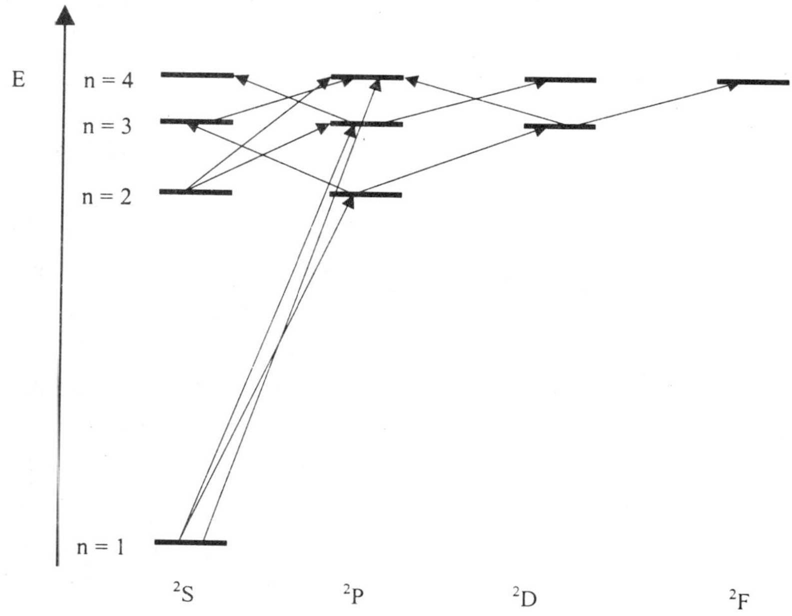

Figure 8-6. Some of the allowed absorption transitions for a hydrogen atom that have been observed are shown.

In terms of term symbols, the selection rules means that since the ground-state of a hydrogen atom is ^2S, the only absorption transition that is allowed is to ^2P states. From an excited ^2P state, transitions to ^2S and ^2D states are allowed. Figure 8-6 shows the energy levels in hydrogen and the observed transitions. In terms of emission spectra, the reverse is also true. Emission and absorption spectra of hydrogen (and other atoms) appear as lines at the various allowed transitions. The specific emission lines in hydrogen for $n_i \rightarrow n_f$ are given by the following expression.

$$\Delta E = h\nu = \frac{1}{2}\left(\frac{1}{n_f^2} - \frac{1}{n_i^2}\right) = \Re_H\left(\frac{1}{n_f^2} - \frac{1}{n_i^2}\right) \qquad (8\text{-}73)$$

The constant term $\Re_H$ is called the Rydberg constant.

Table 8-3. The common emission lines for hydrogen atoms along with the name of the series are listed in the table below.

n_f	n_i	Series	Radiation
1	2, 3, ...	Lyman	ultraviolet
2	3, 4, ...	Balmer	visible
3	4, 5, ...	Paschen	infrared
4	5, 6, ...	Brackett	far infrared
5	6, 7, ...	Pfund	far infrared
6	7, 8, ...	Humphrey's	far infrared

$$\Re_H = \frac{1}{2} hartree = 13.606 eV$$

The various lines seen in an emission spectrum are named by their discoverers or principal investigators and are shown in Table 8-3.

In the case of a hydrogen atom, all of the states are doublets. In the case of multi-electron atoms, there may be different spin states. For light elements where the spin-orbit coupling is weak, the selection rules are as follows.

$$\Delta S = 0 \tag{8-74a}$$

$$|\Delta L| = 1 \tag{8-74b}$$

$$|\Delta J| = 0,1 \tag{8-74c}$$

In addition, the change in L is only allowed if the change comes about in the change in l of only 1 from one electron.

The allowed transitions in atomic spectra will correspond to the strongest transitions. Forbidden transitions will most likely be observed in a spectrum because electron-electron repulsion will result in some mixing of atomic orbitals. Once the assignment of transition lines in an atomic spectrum are made to specific transitions in orbitals, this information can be used to determine such properties as ionization potential, excitation energies, and the extent of spin-orbit coupling.

PROBLEMS AND EXERCISES

8.1) Confirm that a $\psi_{2,1,-1}$ wavefunction for a hydrogen atom is normalized.

8.2) Calculate the average position of the electron from the nucleus for a hydrogen atom in a $2p_{-1}$ orbital. Repeat the calculation for a Li^{+3} ion. How does the average position of the electron compare between an H atom and a Li^{+3} ion?

8.3) Determine how far out the radial coordinate you must integrate in order to capture 90% of the electron density for an electron in a 1s orbital of a hydrogen atom.

8.4) Determine the maximum points and the nodes of a 3s orbital.

8.5) Plot the effective potential in Equation 8-15 for an electron in a d and an f orbital for a hydrogen atom as a function of r. At what point is the potential a minimum for each orbital? How does this compare to the average position for these orbitals?

8.6) Using the following trial wavefunction, determine the ground-state energy of a hydrogen atom using variational theory.

$$\psi_{trial} = Ne^{-\alpha r^2}$$

The term α is an adjustable parameter and N is the normalization constant.

8.7) Using the trial function provided in Equation 8-33, explicitly obtain the expression for the effective nuclear charge ζ in Equation 8-42.

8.8) Determine the STO for an electron in a $2p_0$ orbital in a C atom using the Slater's rules. How does this value compare to the computed values given in Table 8-2?

8.9) Plot the radial component of the STO obtained in Problem 8 as a function of r. On the same plot, also plot the radial component of a $2p_0$ orbital of a one-electron system where Z = 6. What are differences and similarities between the two radial functions?

8.10) Determine the total number of states possible for a Li atom with an electron configuration of $1s^1 2s^1 2p^1$.

8.11) Write the term symbol for an electron in the lowest energy state in the excited Li atom in Problem 10 above.

8.12) Propose an absorption spectrum of a helium atom that is initially in the ground-state. Label each absorption line with the transition using term symbols.

8.13) Calculate the wavelengths for the emission transitions of a He^+ ion as the analog of the Balmer series in a hydrogen atom. What part of the electromagnetic spectrum do these wavelengths correspond to?

8.14) Give the numerical value in atomic units for the following quantities: a) a proton, b) Planck's constant, and c) the speed of light.

Chapter 9

Methods of Molecular Electronic Structure Computations

With the advent of high-speed computers being readily available, electronic structure computations have become an important component of theoretical and experimental chemical research. Calculations may be performed on highly reactive molecules and transition states as reliably as on stable molecules. There are a number of commercially available software with excellent graphics that allow for excellent viewing of three-dimensional structure, electron densities, and dipole moments. Thermodynamic information such as heats of formation and strain energies can be readily obtained from such software. Reaction dynamics can also be obtained such as transition state structures. This chapter will focus on some of the computational details involved in these programs along with a practical "hands-on" guide for using them effectively.

9.1 THE BORN-OPPENHEIMER APPROXIMATION

The Hamiltonian for a molecule is easily determined. The Hamiltonian will include kinetic energy terms for the nuclei (indexed by A) and electrons (indexed by a), electron-nucleus potential (distance of separation r_{Aa}), nuclear-nuclear potential (distance of separation of R_{AB}), and electron-electron repulsion (distance of separation r_{ab}).

$$\hat{H} = -\frac{1}{2}\sum_{A}^{nuclei}\frac{1}{(M_A/m_e)}\nabla_A^2 - \frac{1}{2}\sum_{a}^{electrons}\nabla_a^2 - \sum_{A}^{nuclei}\sum_{a}^{electrons}\frac{Z_A}{r_{Aa}} + \sum_{A>B}^{nuclei}\sum_{B}^{nuclei}\frac{Z_A Z_B}{R_{AB}} + \sum_{a>b}^{electrons}\sum_{b}^{electrons}\frac{1}{r_{ab}} \quad (9\text{-}1)$$

Though the complete Hamiltonian for a molecule is easily determined, the resulting Schroedinger equation is impossible to solve, even analytically.

An approximation that can be made is to realize that the motion of the nuclei is sluggish relative to the motion of the electrons due to the large differences in mass. Due to the great difference in motion between the nuclei and the electrons, the electrons are capable of instantaneously adjusting to any change in position of the nuclei. Hence, the electron motion is determined for a *fixed* position of the nuclei making the distances R_{AB} in Equation 9-1 now a constant. This approximation is called the ***Born-Oppenheimer approximation***. The Born-Oppenheimer approximation removes the kinetic energy operators for the nuclear motion in Equation 9-1.

$$\hat{H} = -\frac{1}{2}\sum_{a}^{electrons}\nabla_a^2 - \sum_{A}^{nuclei}\sum_{a}^{electrons}\frac{Z_A}{r_{Aa}} + \sum_{A>B}^{nuclei}\sum_{B}^{nuclei}\frac{Z_A Z_B}{R_{AB}} + \sum_{a>b}^{electrons}\sum_{b}^{electrons}\frac{1}{r_{ab}} \quad (9\text{-}2)$$

The Schroedinger equation that is solved for then just becomes the electronic Schroedinger equation for the molecule plus a constant term for the nuclear repulsion.

$$\hat{H}^{electronic}\psi^{electronic} = (E^{electronic} + E^{nuclear})\psi^{electronic} \quad (9\text{-}3)$$

The Schroedinger equation is solved for the electrons in a fixed static electric potential arising from the nuclei in that particular arrangement. Different arrangements of the nuclei may then be adopted and the calculation is repeated. The set of solutions obtained can then be used to construct a ***molecular potential energy curve*** for a diatomic molecule or a ***potential energy surface*** for a polyatomic molecule. The lowest point of the potential energy curve or surface is then determined to identify the equilibrium geometry of the molecule (see Figure 9-1). The wavefunctions that result in the computation are called ***molecular orbitals (MO)***.

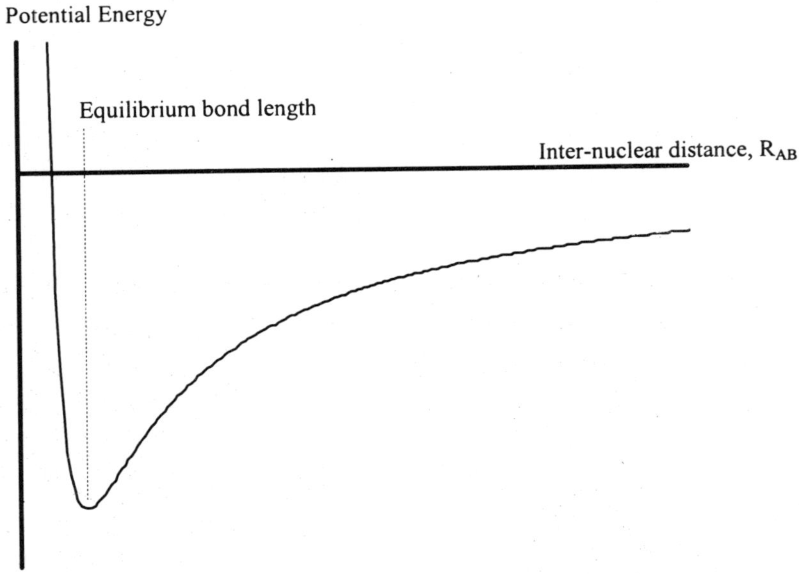

Figure 9-1. The molecular potential energy curve of a diatomic molecule is shown. The minimum point in the curve represents the equilibrium geometry (i.e. equilibrium bond length) of the molecule.

9.2 THE H_2^+ MOLECULE

To obtain an understanding of bonding, it is helpful to first look at the simplest molecular system, H_2^+, where there is only one electron and two nuclei (see Figure 9-2). If the Born-Oppenheimer approximation is made, the Hamiltonian can be readily written from Equation 9-2.

$$\hat{H} = -\frac{1}{2}\nabla^2 - \frac{1}{r_A} - \frac{1}{r_B} + \frac{1}{R_{AB}} \tag{9-4}$$

The term $1/R_{AB}$ is a constant for a particular nuclear configuration. The Schroedinger equation will be only in terms of the electronic motion.

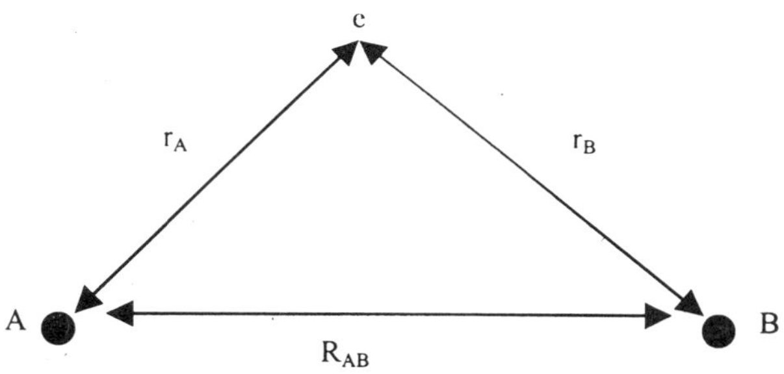

Figure 9-2. The relative positions of the particles in a H_2^+ molecule are shown. In the Born-Oppenheimer approximation, the position of the nuclei is fixed at R_{AB}.

When the electron is close to the nucleus A, then r_B is very large and the potential to nucleus B is negligible making the orbital similar to that of a hydrogen atom. The reverse is also true, when the electron is close to nucleus B, then r_A is very large making the electron appear as a hydrogen-like orbital around the B nucleus. Based on this analysis, it is reasonable to construct the molecular orbitals from a ***linear combination of atomic orbitals (LCAO)***.

$$\psi = \sum_i c_i \varphi_i \tag{9-5}$$

The c_i terms are coefficients indicating the magnitude of contribution of each atomic orbital φ_i to the molecular orbital.

The atomic orbitals used in the sum in Equation 9-5 constitute the basis set. In order to produce a precise molecular orbital, an infinite basis set should be used. Of course in practice only a finite basis set is used. The most severe truncation of this infinite sum is to use the smallest number of functions to hold all of the electrons in an atom and still maintain the spherical symmetry of the atom. This is called a ***minimal basis set***. The minimal basis set of atomic orbitals for the first three periods of the Periodic Table of Elements is given as follows.

H, He:	1s
Li, Be:	1s, 2s
B - Ne:	1s, 2s, $2p_x$, $2p_y$, $2p_z$, 3s
Na, Mg:	1s, 2s, $2p_x$, $2p_y$, $2p_z$, 3s
Al - Ar:	1s, 2s, $2p_x$, $2p_y$, $2p_z$, 3s, $3p_x$, $3p_y$, $3p_z$

For the case of the H_2^+ molecule, the minimal basis set will be two 1s hydrogen atomic orbitals one centered on nucleus A and the other centered on nucleus B.

$$\psi = c_A \varphi_A + c_B \varphi_B \tag{9-6}$$

The optimal values of the coefficients c_A and c_B are determined using Variational theory (see Section 4.1).

$$E = \frac{\langle \psi | \hat{H} | \psi \rangle}{\langle \psi | \psi \rangle} = \frac{\langle c_A \varphi_A + c_B \varphi_B | \hat{H} | c_A \varphi_A + c_B \varphi_B \rangle}{\langle c_A \varphi_A + c_B \varphi_B | c_A \varphi_A + c_B \varphi_B \rangle}$$

$$= \frac{c_A^2 \langle \varphi_A | \hat{H} | \varphi_A \rangle + c_A c_B \langle \varphi_A | \hat{H} | \varphi_B \rangle + c_A c_B \langle \varphi_B | \hat{H} | \varphi_A \rangle + c_B^2 \langle \varphi_B | \hat{H} | \varphi_B \rangle}{c_A^2 \langle \varphi_A | \varphi_A \rangle + c_A c_B \langle \varphi_A | \varphi_B \rangle + c_A c_B \langle \varphi_B | \varphi_A \rangle + c_B^2 \langle \varphi_B | \varphi_B \rangle} \tag{9-7}$$

It is convenient at this point to introduce some shorthand notation. The ***molecular Coulomb integrals*** are symbolized as H_{AA} and H_{BB}. The molecular Coulomb integrals correspond to the Coulombic attraction of the electron to each nucleus.

$$H_{AA} \equiv \langle \varphi_A | \hat{H} | \varphi_A \rangle \tag{9-8}$$

$$H_{BB} \equiv \langle \varphi_B | \hat{H} | \varphi_B \rangle \tag{9-9}$$

The ***resonance integrals*** H_{AB} and H_{BA} have no classical counterpart and are defined as follows.

$$H_{AB} \equiv \langle \varphi_A | \hat{H} | \varphi_B \rangle \tag{9-10}$$

$$H_{BA} \equiv \langle \varphi_B | \hat{H} | \varphi_A \rangle \tag{9-11}$$

Since H_2^+ is a homonuclear diatomic molecule and φ_A and φ_B are real (i.e. hydrogen 1s wavefunctions), $H_{AA} = H_{BB}$ and $H_{AB} = H_{BA}$. The atomic orbitals φ_A and φ_B are normalized resulting in the following simplifications.

$$S_{AA} = \langle \varphi_A | \varphi_A \rangle = 1 \quad \text{and} \quad S_{BB} = \langle \varphi_B | \varphi_B \rangle = 1 \tag{9-12}$$

The last integrals to consider are the ***overlap integrals*** symbolized by S_{AB} and S_{BA}. Since the wavefunctions are real in the ground-state of H_2^+, $S_{AB} = S_{BA}$ and are just symbolized by S.

$$S \equiv \langle \varphi_A | \varphi_B \rangle = \langle \varphi_B | \varphi_A \rangle \tag{9-13}$$

Utilizing the notation and simplifications, the variational energy E in Equation 9-7 can be reduced to the following expression.

$$E = \frac{c_A^2 H_{AA} + 2c_A c_B H_{AB} + c_B^2 H_{BB}}{c_A^2 + 2c_A c_B S + c_B^2} \tag{9-14}$$

Taking the first derivative of E in Equation 9-14 with respect to each coefficient and setting it equal to zero now optimizes the coefficients c_A and c_B resulting in the secular equations.

$$\left(\frac{\partial E}{\partial c_A} \right) c_B = 0$$

$$c_A(H_{AA} - E) + c_B(H_{AB} - SE) = 0 \tag{9-15}$$

$$\left(\frac{\partial E}{\partial c_B} \right) c_A = 0$$

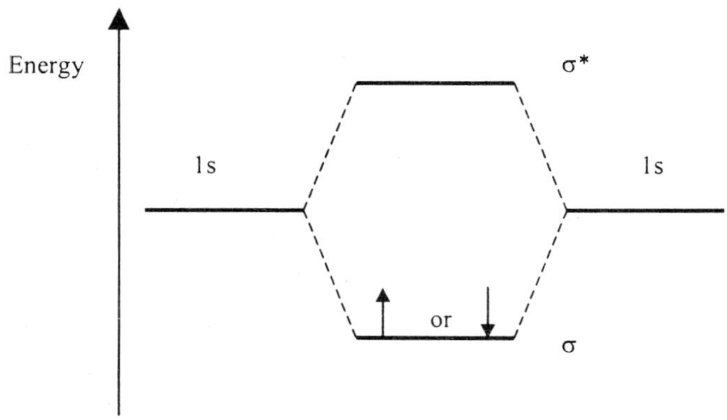

Figure 9-3. The symmetric, σ, bond and the antisymmetric, σ*, are shown for a H_2^+ molecule.

$$c_A(H_{AB} - SE) + c_B(H_{BB} - E) = 0 \tag{9-16}$$

The coefficients are determined from the secular determinant.

$$\begin{vmatrix} H_{AA} - E & H_{AB} - SE \\ H_{AB} - SE & H_{BB} - E \end{vmatrix} = 0 \tag{9-17}$$

Expansion of the secular determinant results in the following expression upon recognition that $H_{AA} = H_{BB}$.

$$(H_{AA} - E)^2 - (H_{AB} - SE)^2 = 0$$

$$(H_{AA} - E) = \pm(H_{AB} - SE) \tag{9-18}$$

The energy E is solved for in Equation 9-18 resulting in two different values. The resulting energy values are shown in Figure 9-3.

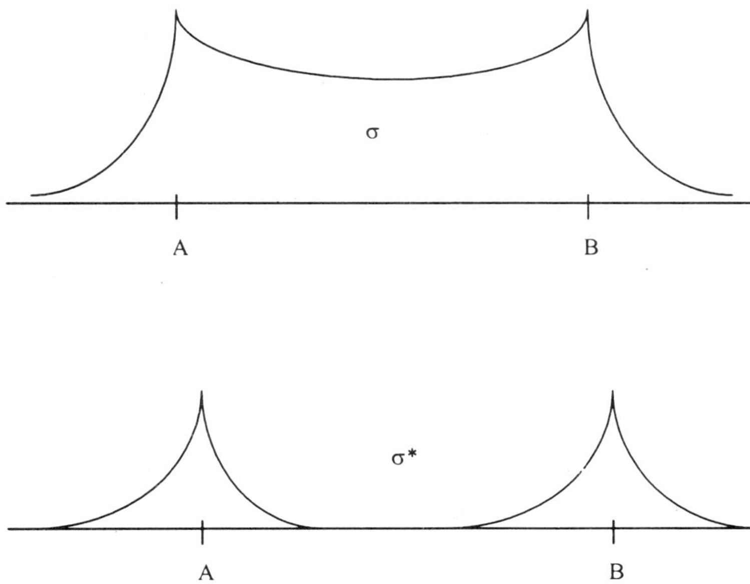

Figure 9-4. The electron densities for the σ and σ* wavefunctions for a H_2^+ molecule are shown. Note that for σ*, there is no electron density between the nuclei and equal electron density on each nucleus.

$$E_\sigma = \frac{H_{AA} + H_{AB}}{1+S} \qquad (9\text{-}19)$$

$$E_{\sigma^*} = \frac{H_{AA} - H_{AB}}{1-S} \qquad (9\text{-}20)$$

The term E_σ is the symmetric bonding mode and E_{σ^*} is the anti-symmetric or anti-bonding mode. Substituting the expressions for E_σ and E_{σ^*} into Equation 9-16 results in the following expressions for the coefficients.

$$c_A = c_B \qquad \text{(symmetric, } \sigma) \qquad (9\text{-}21)$$

$$c_A = -c_B \qquad \text{(anti-symmetric, } \sigma^*) \qquad (9\text{-}22)$$

The wavefunctions for the symmetric and anti-symmetric states can be written as follows.

$$\psi_{\sigma} = c_A(\varphi_A + \varphi_B) \tag{9-23}$$

$$\psi_{\sigma*} = c_A(\varphi_A - \varphi_B) \tag{9-24}$$

The shapes for the wavefunctions are shown in Figure 9-4.

The result for the ground-state of a hydrogen molecule is similar to that for a H_2^+ molecule. There will be an additional electron and it will be placed in the σ orbital of opposite spin with the other electron. When higher order diatomic molecules or excited states of the hydrogen molecule are considered, then p atomic orbitals will combine to form molecular orbitals. When the $2p_z$ atomic orbitals overlap, a σ bond is formed that is symmetric about the inter-nuclear axis. The two $2p_z$ molecular orbital wavefunctions are as follows.

$$\psi_{\sigma}(2p_z) = N(\varphi_{A,2p_z} + \varphi_{B,2p_z}) \tag{9-25}$$

$$\psi_{\sigma*}(2p_z) = N(\varphi_{A,2p_z} - \varphi_{B,2p_z}) \tag{9-26}$$

After the $2p_z$ atomic orbitals form σ bonds, the $2p_x$ and $2p_y$ atomic orbitals combine to form π bonds. The $2p_x$ orbitals form a node in the yz plane, and the $2p_y$ orbitals form a node in the xz plane. Each atomic orbital will form π and π* bonds respectively.

The electrons in a molecule are ordered from the lowest occupied molecular orbital to highest occupied molecular orbital (HOMO). The ordering of electrons occurs in pairs of opposite spins (according to the Pauli principle) as shown in Figure 9-5 for a CO molecule. The ***bond order (BO)*** is determined by taking the sum of electrons in bonding orbitals minus the sum of electrons in anti-bonding orbitals.

Bond order = (# of electrons in bonding orbitals
- # of electrons in anti-bonding orbitals)/2

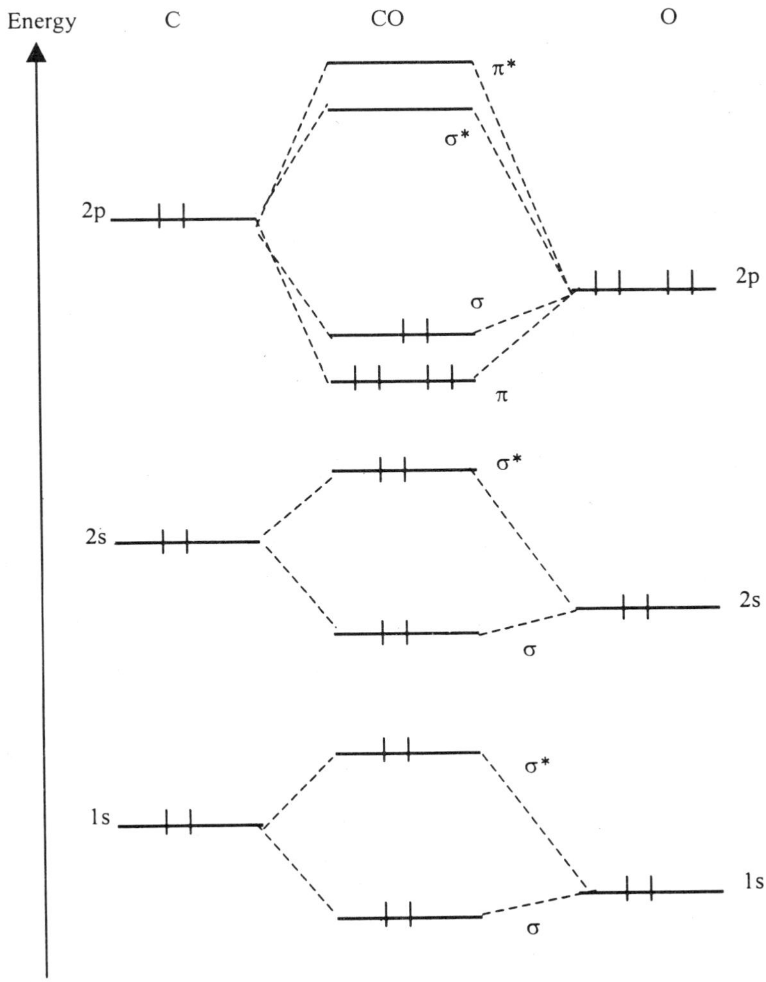

Figure 9-5. The molecular orbitals for a CO molecule are shown. The atomic orbitals for C and O atoms are shown to the left and right and then how they combine to form the molecular orbitals of CO in the center. The vertical lines in each orbital represent an electron.

For the case of CO, there are 10 electrons in bonding orbitals and 4 electrons in anti-bonding orbitals resulting in a bond order of 3; hence, CO is said to have a triple bond. Note that alternatively only electrons in the valence

Table 9-1. The bond orders of a number of diatomic molecules on the basis of LCAO approximation are listed.

	Number of Bonding Electrons	Number of Anti-Bonding Electrons	Bond Order
H_2^+	1	0	1/2
H_2	2	0	1
He_2	2	2	0
Li_2	2	0	1
O_2	8	4	2
Cl_2	8	6	1

orbitals need to be counted to determine the bond order since the electrons in lower levels are filled and will cancel. The bond order for a number of diatomic molecules is shown in Table 9-1.

9.3 MOLECULAR MECHANICS METHODS

In the Born-Oppenheimer approximation, the energy of a molecule is computed for a specific nuclear configuration. An initial "guess" to the nuclear geometry is made and the energy is calculated. The computational cycle is repeated until the equilibrium geometry is obtained. The closer the initial "guess" for the nuclear configuration to the actual equilibrium geometry, the less computational cycles that are needed. As a result, for high-level calculations on large molecules, a good starting point for the computations becomes increasingly more important.

Molecular mechanics methods are a non-quantum (i.e. classical) mechanical computation for obtaining geometries of gas phase molecules. As a result, molecular mechanics methods are computationally fast. Molecular mechanics methods use empirical force fields to describe the energy of a given configuration. The energy of a given configuration is calculated as follows.

$$E = \underbrace{\sum \text{bond stretching}}_{\text{pairs of atoms}} + \underbrace{\sum \text{angle bending}}_{\text{triplets of atoms}} + \underbrace{\sum \text{dihedral}}_{\text{quartets of atoms}}$$

$$+ \sum_{\text{pairs of atoms}} \text{non bonded (van der Waals)} + \sum_{\text{pairs of atoms}} \text{Coulombic}$$

Molecular mechanics treats a molecule as an array of atoms governed by a set of classical-mechanical potential functions. The parameters for the potential functions used in the calculation come from experimental data and/or high-level quantum mechanical computations on similar molecules. The assumption in the technique is that similar bonds in different molecules will have similar properties. This assumption works well so long as the molecules being calculated do not differ significantly from the molecules used to determine the parameters for the force fields. The results that are obtained from the molecular mechanics computation are only as reliable as the empirical data for the force fields. Hooke's Law approximates bond stretching. Angle bending is determined as a given bond angle is deformed from it is optimal angle θ_0 by a form similar to Hooke's law: $k(\theta - \theta_0)^2$. Steric interactions are accounted for by using van der Waals functions that can either be composed of a sixth and twelfth power function or alternatively the twelfth power is replaced with an exponential. It is actually difficult to breakdown the contributions to the molecular potential energy to each separate interaction as many effects are inter-related. So the parameters are spread into each of the different force fields to ensure that experimental results are reproduced.

One very common molecular mechanics package is SYBYL. SYBYL is a simple computation that requires very few data to establish parameters. As a result, SYBYL can be used for elements throughout the Periodic Table. The results of SYBYL of are not as accurate compared to high-level computations, and some results are shown in Tables 9-2 and 9-3.

Another common and more complicated molecular mechanics package is MMFF. MMFF requires much more data to establish the parameters used in the computation. The results are more accurate than SYBYL (as shown in Tables 9-2 and 9-3), but parameters are generally available only for organic molecules and biopolymers.

Molecular mechanics techniques are valuable computations for establishing good starting points of initial geometry for higher-level quantum mechanical computations. The accuracy of the geometries obtained can be

Table 9-2. The calculated geometry using SYBYL and MMFF for a number of different types of bonds are shown. All bond distances are in Å. Data obtained from Wavefunction, Inc. with permission.

bond	molecule	SYBYL	MMFF	experimental
C-C	but-1-yne-3-ene	1.441	1.419	1.431
	propyne	1.458	1.463	1.459
	1,3-butadiene	1.478	1.442	1.483
	propene	1.509	1.493	1.501
	cyclopropane	1.543	1.502	1.510
	propane	1.551	1.520	1.526
	cyclobutane	1.547	1.543	1.548
C=C	cyclopropene	1.317	1.302	1.300
	allene	1.305	1.297	1.308
	propene	1.339	1.339	1.318
	cyclobutene	1.327	1.345	1.332
	but-1-yne-3-ene	1.338	1.337	1.341
	1,3-butadiene	1.338	1.338	1.345
	cyclopentadiene	1.335	1.341	1.345
C≡C	propyne	1.204	1.201	1.206
	but-1-yne-3-ene	1.204	1.200	1.208
C-N	formamide	1.346	1.360	1.376
	trimethylamine	1.483	1.462	1.451
	aziridine	1.484	1.459	1.475
	nitromethane	1.458	1.488	1.489
C-O	formic acid	1.334	1.348	1.343
	dimethyl ether	1.437	1.421	1.410
	methanol	1.437	1.416	1.421
C=O	formic acid	1.220	1.217	1.202
	formaldehyde	1.220	1.225	1.208
	acetone	1.221	1.230	1.222
C-S	dimethylsulfoxide	1.803	1.809	1.799
	dimethylsulfide	1.820	1.808	1.802
	methane thiol	1.821	1.804	1.819
C-Cl	trichloromethane	1.767	1.772	1.758
	dichloromethane	1.767	1.767	1.772
	chloromethane	1.767	1.767	1.781

as close to the values obtained from higher-order quantum mechanical computations. The results from these packages however are generally limited only to geometry and conformational energies.

Table 9-3. Barriers to internal rotation calculated using SYBYL and MMFF for a number of organic molecules are shown. All values are in kcal/mol. Data obtained from Wavefunction, Inc. with permission.

bond	molecule	SYBYL	MMFF	experimental
C-C	CH_3 - CH_3	3.8	4.2	2.88
	CH_3 - CH_2CH_3	4.0	3.4	3.4
	CH_3 - $C(CH_3)_3$	4.7	4.1	4.7
	CH_3 - $CH{=}CH_2$	1.3	2.0	2.0
	CH_3 - CHO	0.0	2.0	1.17
	$H_2C{=}CH$ - $CH{=}CH_2$*	4.9	6.2	5.96
	$H_2C{=}CH$ - CHO	5.2	7.8	7.9
C-N	CH_3 - NH_2	2.0	2.4	1.98
	CH_3 - $NHCH_3$	2.0	3.5	3.62
C-O	CH_3 - OCH_3	4.0	2.4	2.7
	CH_3 - OH	4.0	1.2	1.07
	CH_3 - OCHO	6.9	0.8	1.19

*Barrier relative to trans configuration.

9.4 *AB INITIO* METHODS

The term "*ab initio*" comes from Latin meaning from the beginning. The implication is that the computations are exact with no approximations. This is certainly not the case, as the Schroedinger equation for more than a two-body system cannot be solved without approximations and using approximation techniques. What "*ab initio*" does mean in this context is that the integrals involved in the Schroedinger equation for the system are explicitly solved without the use of empirical parameters.

The first assumption that is made is the Born-Oppenheimer approximation. As described in Section 9.1, this reduces the Schroedinger equation for a molecular system to only the electronic motion for a particular nuclear configuration. As mentioned in Section 9.3, an optimized nuclear configuration as a starting point for an *ab initio* computation can be obtained by using a molecular mechanics method. This reduces the number of computational cycles needed to find the equilibrium geometry and hence energy of the molecule. The Hamiltonian for the molecular system with the Born-Oppenheimer approximation is as given in Equation 9-2.

$$\hat{H} = -\frac{1}{2}\sum_{a}^{electrons}\nabla_a^2 - \sum_{A}^{nuclei}\sum_{a}^{electrons}\frac{Z_A}{r_{Aa}} + \sum_{A>B}^{nuclei}\sum_{B}^{nuclei}\frac{Z_A Z_B}{R_{AB}} + \sum_{a>b}^{electrons}\sum_{b}^{electrons}\frac{1}{r_{ab}}$$

Due to the Born-Oppenheimer approximation, the inter-nuclear distances (the R_{AB} terms) are constant for a particular nuclear configuration. Consequently it is convenient to express the Born-Oppenheimer Hamiltonian as only the operator parts in terms of the electrons, $\hat{H}^{electronic}$, and add the inter-nuclear repulsion potential to the electronic energy, $E^{electronic}$, to obtain the energy of the molecule, E.

$$\hat{H}^{electronic} = -\frac{1}{2}\sum_{a}^{electrons}\nabla_a^2 - \sum_{A}^{nuclei}\sum_{a}^{electrons}\frac{Z_A}{r_{Aa}} + \sum_{a>b}^{electrons}\sum_{b}^{electrons}\frac{1}{r_{ab}} \qquad (9\text{-}27)$$

$$\hat{H}^{electronic}\psi^{electronic} = E^{electronic}\psi^{electronic}$$

$$E = E^{electronic} + \sum_{A>B}^{nuclei}\sum_{B}^{nuclei}\frac{Z_A Z_B}{R_{AB}} \qquad (9\text{-}28)$$

The next step is to make the Hartree-Fock self-consistent field (HF-SCF) approximation as described previously for a multi-electron atom in Section 8.4. The Hartree-Fock approximation results in separation of the electron motions resulting (along with the Pauli principle) in the ordering of the electrons into the molecular orbitals as shown in Figure 9-5 for carbon monoxide. Hence, the many-electron wavefunction ψ for an N-electron molecule is written in terms of one-electron space wavefunctions, f_i, and spin functions, α or β, like what was done for complex atoms in Section 8.4. At this stage it is assumed that the N-electron molecule is a closed-shell molecule (all the electrons are paired in the occupied molecular orbitals). How molecules with open shells are represented will be discussed later in this Section.

$$\psi = f_1(1)\alpha f_1(2)\beta \cdots f_{N/2}(N)\beta \qquad (9\text{-}29)$$

As described in Section 8.4 for a multi-electron atom, the HF-SCF approach assumes that any one electron moves in a potential that is a spherical average due to the other electrons and the nuclei of the molecule.

The potential from the nuclei is set by the initial configuration of the molecule, and the potential from the other electrons are determined from initial approximate wavefunctions resulting in the Hartree-Fock Hamiltonian, $\hat{H}^{eff}$.

$$\hat{H}^{eff}(1) = -\frac{1}{2}\nabla_1^2 - \sum_A^{nuclei}\frac{Z_A}{r_{1A}} + \sum_j^{N/2}[2\hat{J}_j(1) - \hat{K}_j(1)] \qquad (9\text{-}30)$$

The first two terms in Equation 9-30 correspond to the kinetic energy operator of the electron and the attraction between one electron and the nuclei of the molecule. These first two terms constitute what is called the core Hamiltonian - no interactions from other electrons (see Equation 8-52). The next term, $\hat{J}_j(1)$, is the ***Coulomb operator***.

$$\hat{J}_j(1) = \int |f_j(2)|^2 \frac{1}{r_{12}} d\tau_2 \qquad (9\text{-}31)$$

The Coulomb operator accounts for the smeared-out electron potential with an electron density of $|f_j(2)|^2$ (the factor of 2 arises because there are two electrons in each spatial orbital). The last term in Equation 9-30 is the ***exchange operator.***

$$\hat{K}_j(1) = \int \frac{f_j^*(2) f_i(2)}{r_{12}} d\tau_2 \qquad (9\text{-}32)$$

The exchange operator has no physical interpretation as it takes into account the effects of spin correlation.

The Schroedinger equation is now solved for the one electron, $f_i(1)$.

$$\hat{H}^{eff}(1) f_i(1) = \varepsilon_i f_i(1) \qquad (9\text{-}33)$$

The term ε_i corresponds to the orbital energy of the electron ascribed by $f_i(1)$. The molecular orbital wavefunctions, f_i, are eigenfunctions of the Hartree-Fock Hamiltonian operator, $\hat{H}^{eff}$, and can be chosen to be orthogonal causing many integrals in the expression to vanish.

The true Hamiltonian and wavefunction of a molecule includes the coordinates of all N electrons. The Hartree-Fock Hamiltonian includes the coordinates of only one electron and is a differential equation in terms of only one electron. As can be seen by Equations 9-31 and 9-32, the Hartree-Fock Hamiltonian depends on its eigenfunctions that must be known before Equation 9-33 can be solved. As in the case for multi-electron atoms in Section 8.4, the solution of the Hartree-Fock equations must be done in an iterative process. The energy of the molecule in terms of the Hartree-Fock approach, E^{HF}, is determined as follows.

$$E^{HF} = 2\sum_{i}^{\frac{N}{2}} \varepsilon_i - \sum_{i}^{\frac{N}{2}}\sum_{j}^{\frac{N}{2}} (2J_{ij} - K_{ij}) + \sum_{A>B}^{nuclei}\sum_{B}^{nuclei} \frac{Z_A Z_B}{R_{AB}} \tag{9-34}$$

The first summation in Equation 9-34 is over all the orbital energies of the occupied molecular orbitals (again, the factor 2 is needed because there are two electrons in each molecular orbital). The terms J_{ij} and K_{ij} are determined by operating the Coulomb operator (Equation 9-31) and the exchange operator (Equation 9-32) on $f_i(1)$ and multiplying the result by $f_i^*(1)$ and integrating overall space. The last summation term in Equation 9-34 refers to the inter-nuclear repuslion potential for a particular nuclear configuration.

The spatial one-electron wavefunctions, $f_i(n)$, are represented as a linear combination of atom-centered functions (i.e. atomic orbitals), φ_{ik}, called the linear combination of atomic orbitals (LCAO) approximation. The functions φ_{ik} constitute a basis set. This is the same approach used for multi-electron atoms and for the H_2^+ molecule. The index k refers to the specific atomic orbital wavefunction, and the index i refers to its contribution to a specific molecular orbital.

$$f_i(n) = \sum_{k}^{N'} c_{ik}\varphi_{ik} \tag{9-35}$$

The best representation of the molecular orbital occurs when an infinite sum of atomic orbitals is made, but of course in practice only a finite N' sum is used. (Note that the term N' should not be confused with N corresponding to the total number of electrons in the molecule). The coefficients c_{ik} correspond to the contribution of each atomic orbital to the corresponding molecular orbital.

The energy of a given electron in a molecular orbital of the molecule, ε_i, is calculated as a function of the coefficients for that molecular orbital, c_{ik}. These equations are called the ***Roothaan-Hall equations***. Equation 9-35 is substituted into Equation 9-33.

$$\sum_k c_{ik} \hat{H}^{eff} \varphi_k = \varepsilon_i \sum_k c_{ik} \varphi_k \tag{9-36}$$

In order to calculate $\hat{H}^{eff}$, an initial "guess" to the coefficients for the other molecular orbitals f_j must be made. Multiplying Equation 9-36 by φ_j* (where $j = 1, 2, 3,..., N'$) and integrating yields the following expression.

$$\sum_k c_{ik} (H_{jk}^{eff} - \varepsilon_i S_{jk}) = 0 \tag{9-37}$$

The terms H_{jk}^{eff} are called the ***Fock matrix***.

$$H_{jk}^{eff} = \langle \varphi_j | \hat{H}^{eff} | \varphi_i \rangle \tag{9-38}$$

The terms S_{jk} are the ***overlap matrix***.

$$S_{jk} = \langle \varphi_j | \varphi_k \rangle \tag{9-39}$$

Using Variational theory, the coefficients are optimized by taking the derivative of ε_i with respect to each coefficient and setting it equal to zero. This results in a set of equations similar to that obtained for the H_2^+ molecule (Equations 9-15 and 9-16).

$$\det\left(H_{jk}^{eff} - \varepsilon_i S_{jk}\right) = 0 \tag{9-40}$$

The optimized coefficients obtained from Equation 9-40 for each molecular orbital in turn are then compared to the initial "guess" for the coefficients. If there is a differance, the computation is repeated with the new optimized coefficients. If there is no significant difference or enough computational cycles have been completed so that there is no significant difference, the computation is terminated. This iterative process (as in the case for atoms as described in Chapter 8) is called a self-consistent field.

Point of Further Understanding

Describe the similarities and differences between the HF-SCF-LCAO computation of a multi-electron molecule and that for a multi-electron atom as described in Section 8.4 and for the H_2^+ molecule described in Section 9.2.

In the case of a closed-shell molecule where all of the electrons in the occupied molecular orbitals are paired, the wavefunction representation is as described in Equation 9-29. Numerous cancellations will occur when integrating over these spin wavefunctions due to the orthonormality of the spin functions α and β. Electrons with like spins will interact and electrons with unlike spins will not interact. The function represented in Equation 9-29 is termed a ***restricted Hartree-Fock (RHF)*** wavefunction. An example of a closed shell molecule is carbon monoxide. There are a total of 14 electrons that are paired up in seven molecular orbitals (see Figure 9-5).

There are two commonly used procedures for open-shell molecules. In an open-shell molecule, not all of the electrons in the molecular orbitals are paired. An example of an open-shell molecule is nitrogen monoxide. There are a total of 15 electrons occupying eight molecular orbitals. One procedure for open-shell molecules is to use an RHF wavefunction as in a closed-shell molecule. The difficulty with this approach is that the lone electron in the molecular orbital will interact only with the other electrons in the molecule with the same spin. To relax this constraint on the solution, each electron in a molecular orbital is given a different spatial function. The relaxing of the constraint that electrons must occupy molecular orbitals in pairs is called the ***unrestricted Hartree-Fock (UHF)*** wavefunctions. Variational energies calculated using UHF wavefunctions are generally lower in energy than those calculated using RHF wavefunctions. One difficulty, however, with UHF wavefunctions is that they may not be eigenfunctions of the total squared spin angular momentum operator, $\hat{S}^2$, whereas RHF wavefunctions are eigenfunctions. This can lead to impure spin states for the molecule. In practice, the S^2 expectation value is calculated for a UHF wavefunction. If the value results in the true value of $\hbar^2 S(S + 1)$, the UHF wavefunction is a reasonable molecular wavefunction. Often though the UHF wavefunction is used as a first approximation to the true molecular wavefunction even if a significant discrepancy exists due to the lower Variational energy obtained.

The next thing to be specified is the kind of functions, φ_{ik}, to be used for the LCAO approximation in Equation 9-35.

$$f_i(n) = \sum_k^{N'} c_{ik} \varphi_{ik}$$

From the experience of the He atom computation in Section 8.2, the hydrogen atom wavefunctions are not a good choice as a starting point for systems with more than one electron due to electron shielding effects. One reasonable choice for a set of basis functions is Slater-type orbitals (STO's) as introduced in Section 8.4. In practice though, most molecular HF-SCF-LCAO computations use ***Gaussian-type orbitals (GTO)***. A Gaussian function centered on nucleus A has the following form.

$$N r^l e^{-\varsigma \cdot r_A^2} Y_l^m(\theta, \phi) \qquad (9\text{-}41)$$

The term N is the normalization constant, and ζ is the shielding constant. The radial part of the Gaussian function is similar to that used for the harmonic oscillator wavefunctions. The general shapes are shown in Figure 5-1 where the origin for the Gaussian function will be at nucleus A. As with STO's, the spherical harmonics, $Y_l^m(\theta, \phi)$, are used in conjunction with the radial component. The Gaussian functions do not have the proper cusp at the nucleus (i.e. small values of r_A) for atomic orbitals; hence atomic orbitals are represented by a linear combination of the Gaussian functions. This results in an increase in the number of integrals that must be solved in a HF-SCF-LCAO computation; however, the computer computational time is reduced for GTO's than for STO's. The reason for the decreased computer computational time is because two Gaussian functions centered at two different nuclei is equal to a single Gaussian centered at a third point.

The number of functions, N', to be used in the LCAO approximation now needs to be determined. As mentioned in Section 9.2, the smallest number of functions that can be used is the minimal basis set. The minimal basis set

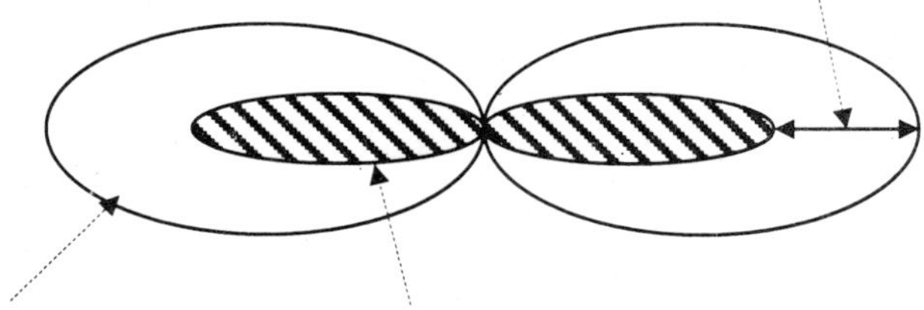

Figure 9-6. A schematic of the effect of a split-valence p-orbital is shown. The size of the atomic orbital in its contribution to the molecular orbital can be varied within the limits set by the inner and outer functions.

is comprised of the minimum number of atomic orbitals needed to hold all of the electrons in a given atom. There are two shortcomings with using a minimal basis set for molecular systems. One shortcoming is that all basis functions are either themselves spherical (such as s functions) or come in sets that describe a sphere (such as p functions). Consequently, molecules that incorporate only atoms with a spherical environment are better described by a minimal basis set than molecules that incorporate atoms with an aspherical environment. The other shortcoming of a minimal basis set is that the basis functions are atom centered. This restricts the flexibility of the functions to describe electron distribution between the nuclei to form chemical bonds. The net effect of these shortcomings is to make molecules too ionic and bonds too long. The obvious answer to both shortcomings is to add more functions to the basis set. An increased basis set means that there are more adjustable parameters in the Variational optimization; however, it comes at the expense of increased computational effort.

The first shortcoming of a minimal basis set, that the basis functions are too spherical, can be resolved by introducing a ***split-valence basis set***. In these basis sets, the valence atomic orbitals are split into two parts: an inner compact orbital, and an outer more diffuse one. The coefficients of the inner and outer orbitals can be varied independently in the construction of the

Table 9-4. The HF-SCF-LCAO calculated geometry for a number of different types of bonds using different basis sets. All bond distances are in Å. Data obtained from Wavefunction, Inc. with permission.

bond	molecule	STO-3G	3-21G	6-31G*	experiment
C-C	but-1-yne-3-ene	1.459	1.432	1.439	1.431
	propyne	1.484	1.466	1.468	1.459
	1,3-butadiene	1.488	1.479	1.468	1.483
	propene	1.520	1.510	1.503	1.501
	cyclopropane	1.502	1.513	1.497	1.510
	propane	1.541	1.541	1.528	1.526
	cyclobutane	1.554	1.571	1.548	1.548
C=C	cyclopropene	1.277	1.282	1.276	1.300
	allene	1.288	1.292	1.296	1.308
	popene	1.308	1.316	1.318	1.318
	cyclobutene	1.314	1.326	1.322	1.332
	but-1-yne-3-ene	1.320	1.320	1.322	1.341
	1,3-butadiene	1.313	1.320	1.323	1.345
C≡C	propyne	1.170	1.188	1.187	1.206
	but-1-yne-3-ene	1.171	1.190	1.188	1.208
C-O	formic acid	1.385	1.350	1.323	1.343
	dimethyl ether	1.433	1.433	1.391	1.410
C=O	formic acid	1.214	1.198	1.182	1.202
	formaldehyde	1.217	1.207	1.184	1.208
	acetone	1.219	1.211	1.192	1.222

molecular orbital in the SCF computation. The size of the atomic orbital that contributes to the molecular orbital can be varied within the limits set by the inner and outer basis functions as shown in Figure 9-6 for a p-orbital. One common type of split-valence basis set used is a 3-21G. This nomenclature means that the core orbitals (the first number before the dash) are made up of 3 Gaussian functions. The inner valence orbitals are made up of 2 Gaussian functions (next number after the dash), and the outer valence orbitals is made up of 1 Gaussian function.

Table 9-5. Bond dissociation energy ($HX \rightarrow H^+ + X^-$) is shown calculated using HF-SCF-LCAO with different basis sets. All values are expressed in kcal/mol. Data obtained from Wavefunction, Inc. with permission.

HX	3-21G$^{(*)}$	6-31G*	experimental
CH_4	463	457	426
NH_3	463	444	409
H_2O	450	429	398
C_2H_2	405	403	381
SiH_4	390	388	378
HF	432	409	376
PH_3	387	383	376
H_2S	364	360	358
HCN	379	370	359
HCl	337	335	337

Molecules that contain the atoms Li and Be need to have 2p functions, and Na and Mg need 3p functions added to the basis set in order to prevent the compounds from becoming too ionic. Likewise, elements in the second-row and heavier main group elements of the Periodic Table need d-type functions added to the basis set. For second-row elements, the split-valence basis set 3-21G$^{(*)}$ is used. The "$^{(*)}$" in the nomenclature indicates that d-type orbitals are available for second-row elements only.

The next shortcoming of a minimal basis set, that the atomic orbitals are atom centered, must now be resolved. This could be resolved by adding functions that are off-centered from the nuclei. This is a dangerous solution however, because it becomes easy to bias the result. A better approach is to continue on the philosophy that more atomic orbital functions in the sum improve the result. The solution is to add p-type functions on hydrogen and d-type functions on main-group heavy atoms that allow the displacement of the electron density away from the nuclear positions. These types of basis sets are called ***polarization basis sets***. Examples of polarization basis sets include 6-31G* and 6-31G**. In a 6-31G* basis set, d-type orbitals are added to heavy main group elements. In a 6-31G** basis set, p-type orbitals are added to hydrogen along with the d-type orbitals in heavy main group elements.

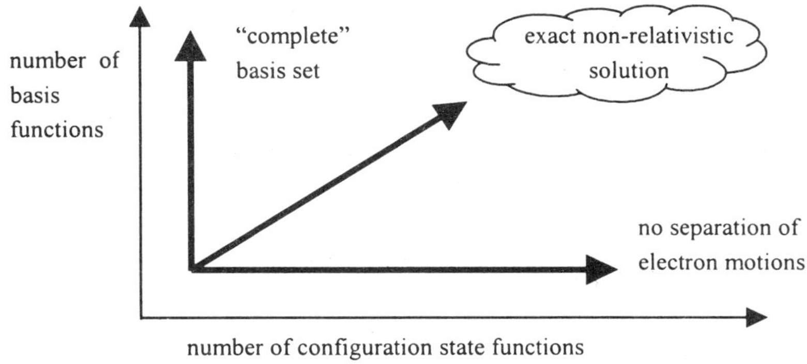

Figure 9-7. The approach to an exact solution of a system with multiple electrons as the number of basis functions and as the configuration state functions is increased is depicted in this diagram.

The result from a HF-SCF-LCAO computation includes information on the equilibrium geometry of the molecule in addition to thermodynamic information such as total energy of the molecule, heats of formation, and bond dissociation energy. The results of some HF-SCF-LCAO computations are shown in Tables 9-4 and 9-5.

Increasingly larger basis sets can be used in HF-SCF-LCAO computations at the expense of computation time; however, the 6-31G** basis set represents a practical limit for these types of computations on medium sized molecules. This is because the HF approximation of separation of electron motion becomes increasingly more important. As seen in Table 9-5, the dissociation energy obtained from HF-SCF-LCAO computations is not very good relative to exerimental values. The problem is due to electron correlation. To gain some understanding of the correlation problem, consider the ground-state of a helium atom. Both electrons are in the 1s state. Since the electrons repel, they tend to stay away from each other. If one electron is close to the nucleus at a given instant, it is energetically more favorable for the other electron to be far away from the nucleus. The problem comes about because the HF approximation solves for the wavefunction of a particular electron with respect to an average charge distribution from the other electrons without allowing for instantaneous

Table 9-6. The MP2 calculated geometries for a number of different types of bonds using different basis is shown. All bond distances are expressed in Å. Data obtained from Wavefunction, Inc. with permission.

bond	molecule	MP2/6-31G*	MP2/6-311+G(2d,p)	experiment
C-C	but-1-yne-3-ene	1.429	1.429	1.431
	propyne	1.463	1.463	1.459
	1,3 butadiene	1.458	1.456	1.483
	propene	1.499	1.499	1.501
	cyclopropane	1.504	1.508	1.510
	propane	1.526	1.526	1.526
	cyclobutane	1.545	1.549	1.548
C=C	cyclopropene	1.303	1.299	1.300
	allene	1.313	1.310	1.308
	propene	1.338	1.336	1.318
	cyclobutene	1.347	1.346	1.332
	but-1-yne-3-ene	1.344	1.342	1.341
	1,3-butadiene	1.344	1.342	1.345
	cyclopentadiene	1.354	1.354	1.345
C≡C	propyne	1.220	1.214	1.206
	but-1-yne-3-ene	1.223	1.217	1.208
C-O	formic acid	1.351	1.349	1.343
	dimethyl ether	1.416	1.415	1.410
C=O	formic acid	1.214	1.205	1.202
	formaldehyde	1.221	1.213	1.208
	acetone	1.228	1.220	1.222

adjustment of that particular electron's trajectories as a result of the other electrons. The approximation of separation of electron motion can be relaxed by using ***correlated models***. The effect of increasing the number of functions added to the LCAO and the relaxation of the HF approximation of electron separation to the true wavefunction of a multi-electron system is shown diagramatically in Figure 9-7. To correct the electron correlation problem, there are two different correlated models: ***Configuration Interaction (CI)*** and ***Moller-Plesset (MPn)*** methods.

In the CI methods, the electron correlation is considered by taking a linear combination of the HF ground-state wavefunction with a large number of excited configurations. The expansion coefficients are then varied using a Variational approach until a minimum energy is achieved. Since excited configurations have a large percentage of their probability density far away from the nuclei, convergence is slow and large numbers of configurations

Table 9-7. Bond dissociation energy (HX → H^+ + X^-) is shown calculated using MP2 with different basis sets. All values are expressed in kcal/mol. Data obtained from Wavefunction, Inc. with permission.

HX	MP2/6-31G*	MP2/6-311+G(2d,p)	experimental
CH_4	458	426	426
NH_3	444	412	409
H_2O	429	396	398
C_2H_2	403	382	381
SiH_4	386	382	378
HF	411	374	376
PH_3	383	375	376
H_2S	363	356	358
HCN	374	355	359
HCl	339	335	337

must be included. In practical CI methods, transitions of only the electrons in the highest occupied molecular orbital (HOMO) to the lowest unoccupied molecular orbital (LUMO) are considered. The level to which this is done is prescribed by the particular method chosen. Some the common ones include the following: ***Configuration Interaction Single excitations (CIS)*** only, ***Configuration Interaction Double excitations (CID)*** only, and ***Configuration Interaction Single and Double excitations (CISD)*** only.

The Moller-Plesset method uses perturbation theory to correct for the electron correlation in a many-electron system. The Moller-Plesset method has the advantage that it is a computationally faster approach than CI computations; however, the disadvantage is that it is not Variational. A non-Variational result is not, in general, an upper bound of the true ground-state energy. In the Moller-Plesset method, the zero-order Hamiltonian is defined as the sum of all the N one-electron Hartree-Fock Hamiltonians, $\hat{H}_i^{HF}$, as given in Equation 9-30.

$$\hat{H}^{(0)} = \sum_{i=1}^{N} \hat{H}_i^{HF} \tag{9-42}$$

The first-order perturbation is the difference between the zero-order Hamiltonian in Equation 9-42 and the electronic Hamiltonian in Equation 9-27.

$$\hat{H}^{(1)} = \hat{H}^{electronic} - \hat{H}^{(0)} \tag{9-43}$$

The Hartree-Fock ground-state wavefunction (Equation 9-29), ψ^{HF}, is an eigenfunction of the Hartree-Fock Hamiltonian, $\hat{H}^{(0)}$, with an eigenvalue of $E^{(0)}$ (the sum of the orbital energies of all the occupied spin-orbitals). The HF energy associated with the normalized ground-state HF wavefunction is the following expectation value.

$$\begin{aligned} E_{HF} &= \langle \psi^{HF} | \hat{H}^{electronic} | \psi^{HF} \rangle = \langle \psi^{HF} | \hat{H}^{(0)} | \psi^{HF} \rangle + \langle \psi^{HF} | \hat{H}^{(1)} | \psi^{HF} \rangle \\ &= E^{(0)} + E^{(1)} \end{aligned} \tag{9-44}$$

Hence, the HF energy is the sum of the zero and first-order energy. The first correction to the ground-state energy of the system as a result of electron correlation is given by second-order perturbation theory.

$$E_0^{(2)} = \sum_{J \neq 0} \frac{\langle \psi_J^{HF} | \hat{H}^{(1)} | \psi_0^{HF} \rangle \langle \psi_0^{HF} | \hat{H}^{(1)} | \psi_J^{HF} \rangle}{E_0^{(0)} - E_J} \tag{9-45}$$

A Moller-Plesset computation to a second-order energy correction is called an ***MP2*** computation, and higher-order energy corrections are called ***MP3***, ***MP4***, and so on.

Some results from MP2 computations are shown in Tables 9-6 and 9-7. As can be seen in Table 9-6, the bond distances obtained from MP2 computations, in general, are in good agreement with experiment; however, the bond distances in multiple bonds such as C=C and C≡C are not as good compared to experiment as other methods such as HF-LCAO-SCF. There is however significant improvement in thermodynamic quantities such as bond dissociation energies. The most significant advantage of using correlated models is to obtain reliable thermodynamic information.

9.5 SEMI-EMPIRICAL METHODS

The *ab initio* methods, especially when including electron correlation, are computationally intensive. This limits their ability to handle large sized molecules. Semi-empirical methods provide a means for obtaining computational results for large sized molecules and inorganic molecules including transition state elements. Semi-empirical methods employ some of the same elements as *ab initio* computations except some integrals are ignored and replaced with parameters to reproduce experimental results. Semi-empirical methods can be thought of as a blend of the *ab initio* methods and molecular mechanics.

Semi-empirical models begin with the HF and LCAO approximations resulting in the Roothaan-Hall equations (Equations 9-36 through 9-40). A minimal basis set is used of STO's. The Roothaan-Hall equations are solved in a self-consistent field fashion, however not all of the integrals are actually solved. In the most severe approximation, there is ***complete neglect of differential overlap (CNDO)***.

$$S_{jk} = \langle \varphi_j | \varphi_k \rangle$$

In the CNDO approximation, the value of this integral is taken to be zero even when different atomic orbitals belong to the same atom. The surviving integrals from the Roothaan-Hall equations are often taken as parameters with values that are adjusted until the results from the CNDO computation resemble those of Hartree-Fock SCF minimal basis set computations. Less severe truncations have been developed called ***modified neglect of differential overlap (MNDO)***. In MNDO, only the differential overlap is ignored when the basis functions belong to different atoms. Several other improvements have been made in packages that include ***Austin Model 1 (AM1)*** and ***PM3***.

The values of the parameters used in semi-empirical computations, so that their results agree with experiment and that of HF-SCF-LCAO computations with minimal basis sets, must come from either experimental values or computational values much like in the case of molecular mechanics. As a consequence, care must be taken to use these packages only for the type of molecules for which the packages have been parameterized. The MNDO

Table 9-8. Bond distances calculated using AM1 and PM3 semi-empirical methods. All bond distances are in Å. Data obtained from Wavefunction, Inc. with permission.

bond	molecule	AM1	PM3	experiment
C-C	but-1-yne-3-ene	1.405	1.414	1.431
	propyne	1.427	1.432	1.459
	1,3-butadiene	1.451	1.456	1.483
	propene	1.476	1.480	1.501
	cyclopropane	1.501	1.499	1.510
	propane	1.507	1.512	1.526
	cyclobutane	1.543	1.542	1.548
C=C	cyclopropene	1.318	1.314	1.300
	allene	1.298	1.297	1.308
	propene	1.331	1.328	1.318
	cyclobutene	1.354	1.349	1.332
	but-1-yne-3-ene	1.336	1.332	1.341
	1,3-butadiene	1.335	1.331	1.345
C≡C	propyne	1.197	1.191	1.206
	but-1-yne-3-ene	1.198	1.193	1.208
C-O	formic acid	1.357	1.344	1.343
	dimethyl ether	1.417	1.406	1.410
C=O	formic acid	1.230	1.211	1.202
	formaldehyde	1.227	1.202	1.208
	acetone	1.235	1.217	1.222

and AM1 models have been parameterized primarily for organic molecules. The PM3 model has been parameterized for organic molecules and certain transition metals listed below.

Ti, Cr, Mn, Fe, Co, Ni, Cu, Zn, Zr, Mo, Ru, Rh, Pd, Cd, Hf, Ta, W, Hg

The parameters for MNDO, AM1, and PM3 have been parameterized to reproduce experimental equilibrium geometries and heats of formation of organic compounds. The parameters for the PM3 model on transition metals have been determined solely on the basis of reproducing equilibrium geometries of transition metal inorganic compounds and organometallics. The lack of PM3 parameterization to reproduce thermochemical information about transition metal inorganics and organometallics is as a result of a general lack of thermochemical information on these compounds.

Table 9-9. Bond dissociation energy ($HX \rightarrow H^+ + X^-$) is shown calculated using AM1 and PM3 semi-empirical methods. All values are expressed in kcal/mol. Data obtained from Wavefunction, Inc. with permission.

HX	AM1	PM3	experimental
CH_4	434	432	426
NH_3	427	408	409
H_2O	412	403	398
C_2H_2	401	392	381
SiH_4	361	352	378
HF	445	399	376
PH_3	347	352	376
H_2S	380	362	358
HCN	354	347	359
HCl	354	336	337

Results from semi-empirical computations are shown in Tables 9-8 and 9-9. The geometrical information from these computations is in good agreement with experimental information; however, the thermochemical information, in general, is not in good agreement. Semi-empirical thermochemical computational data is generally not accurate enough for absolute values; however, it is useful for comparison purposes to explain or predict trends.

9.6 DENSITY FUNCTIONAL METHODS

All *ab initio* methods start with a Hartree-Fock (HF) approximation that result in the spin orbitals, and then electron correlation is taken into account. Though the results of such calculations are reliable, the major disadvantage is that they are computationally intensive and cannot be readily applied to large molecules of interest. ***Density functional (DF)*** methods provide an alternative route that, in general, provide results comparable to CI and MP2 computational results; however, the difference is that DF computations can be done on molecules with 100 or more heavy atoms.

Table 9-10. Bond distances calculated using density functional methods. All bond distances are in Å. Data obtained from Wavefunction, Inc. with permission.

bond	molecule	SVWDN/DN*	pBP/DN*	pBP/DN**	experiment
C-C	but-1-yne-3-ene	1.406	1.422	1.422	1.431
	propyne	1.438	1.456	1.456	1.459
	1,3-butadiene	1.437	1.455	1.456	1.483
	propene	1.479	1.500	1.501	1.501
	cyclopropane	1.493	1.509	1.511	1.510
	propane	1.508	1.538	1.541	1.526
	cyclobutane	1.538	1.556	1.558	1.548
C=C	cyclopropene	1.292	1.300	1.301	1.300
	allene	1.303	1.313	1.313	1.308
	propene	1.330	1.342	1.343	1.318
	cyclobutene	1.340	1.349	1.350	1.332
	but-1-yne-3-ene	1.338	1.350	1.351	1.341
	1,3-butadiene	1.338	1.349	1.350	1.345
C≡C	propyne	1.210	1.216	1.217	1.206
	but-1-yne-3-ene	1.213	1.220	1.220	1.208
C-O	formic acid	1.336	1.359	1.359	1.343
	dimethyl ether	1.393	1.424	1.426	1.410
C=O	formic acid	1.202	1.211	1.212	1.202
	formaldehyde	1.203	1.212	1.213	1.208
	acetone	1.215	1.224	1.244	1.222

In HF models, the computation begins with an exact Hamiltonian but an approximate wavefunction written as a product of one-electron functions. The solution is improved by optimizing the one-electron functions (the value and number of coefficients in the LCAO approximation) and by increasing the flexibility of the final wavefunction representation (electron correlation). By contrast, DF models start with a Hamiltonian corresponding to an "idealized" many-electron system for which an exact wavefunction is known. The solution is obtained by optimizing the "ideal" system closer and closer to the real system.

In HF models, the energy of the system, E^{HF}, (see Equation 9-34) is written as follows.

$$E^{HF} = E^{core} + E^{nuclear} + E^{Coulomb} + E^{exhange} \qquad (9\text{-}46)$$

Table 9-11. Hydrogenation energies for a number of different organic reactions calculated using density functional methods are shown. All energies are expressed in kcal/mol. Data obtained from Wavefunction, Inc. with permission.

reaction	SVDN/DN*	pBP/DN*	pBP/DN**	experiment
$C_2H_6 + H_2 \rightarrow 2CH_4$	-16	-18	-19	-19
$CH_3NH_2 + H_2 \rightarrow CH_4 + NH_3$	-23	-24	-27	-26
$CH_3OH + H_2 \rightarrow CH_4 + H_2O$	-26	-26	-30	-30
$CH_3F + H_2 \rightarrow CH_4 + HF$	-26	-25	-28	-29
$F_2 + H_2 \rightarrow 2HF$	-126	-119	-126	-133
$C_2H_4 + H_2 \rightarrow 2CH_4$	-66	-57	-58	-57
$C_2H_2 + 3H_2 \rightarrow 2CH_4$	-126	-107	-109	-105

The E^{core} is the energy of the single electron with the nucleus. The $E^{nuclear}$ energy is the repulsion between the nuclei for a given nuclear configuration. The term $E^{Coulomb}$ is the energy of repulsion between the electrons. The last term, $E^{exhange}$, takes the spin-correlation into account. In DF models, the energy of the system is comprised of the same core, nuclear, and Coulomb parts, but the exchange energy along with the correlation energy, $E_{XC}(\rho)$, is accounted for in terms of a function of the electron density matrix, ρ(r).

$$E^{DF} = E^{core} + E^{nuclear} + E^{Coulomb} + E_{XC}[\rho] \qquad (9\text{-}47)$$

In the simplest approach, called ***local density functional theory***, the exchange and correlation energy are determined as an integral of some function of the total electron density.

$$E_{XC} = \int \rho(r)\varepsilon_{XC}[\rho(r)]dr \qquad (9\text{-}48)$$

The electron density matrix, ρ(r), is determined from the ***Kohn-Sham orbitals***, ψ_i, as given in the following expression for a system with N electrons.

$$\rho(r) = \sum_{i=1}^{N} \left|\psi_i\right|^2 \qquad (9\text{-}49)$$

The term $\varepsilon_{XC}[\rho(r)]$ is the exchange-correlation energy per electron in a homogeneous electron gas of constant density.

The Kohn-Sham wavefunctions are determined from the ***Kohn-Sham equations***. The following expression is for a system of N-electrons.

$$\left\{-\frac{1}{2}\nabla_1^2 - \sum_A^{nuclei}\frac{Z_A}{r_{A,1}} + \int\frac{\rho(r_2)}{r_{12}}dr_2 + V_{XC}(r_1)\right\}\psi_1(r_1) = \varepsilon_i\psi_i(r_1) \qquad (9\text{-}50)$$

The terms ε_i are the Kohn-Sham orbital energies. The correlation exchange potential, V_{XC}, is the functional derivative of the exchange-correlation energy.

$$V_{XC}[\rho] = \frac{\delta E_{XC}[\rho]}{\delta\rho} \qquad (9\text{-}51)$$

If E_{XC} is known, then V_{XC} can be computed.

The Kohn-Sham equations are solved in a self-consistent field fashion. Initially a charge density is needed so that E_{XC} can be computed. To obtain the charge density, an initial "guess" to the Kohn-Sham orbitals is needed. This initial guess can be obtained from a set of basis functions whereby the coefficients of expansion of the basis functions can be optimized just like in the HF method. From the function of E_{XC} in terms of the density, the term V_{XC} is computed. The Kohn-Sham equations (Equation 9-50) are then solved to obtain an improved set of Kohn-Sham orbitals. The improved set of Kohn-Sham orbitals is then used to calculate a better density. This iterative process is repeated until the exchange-correlation energy and the density converge to within some tolerance.

A common type of local density functional Hamiltonian is the ***SVWN***. The local density functional theory represents a severe approximation for molecular systems since it assumes a uniform total electron density throughout the molecular system. Other approaches have been developed that account for variation in total density (non-local density functional theory). This is done by having the functions depend explicitly on the gradient of the density in addition to the density itself. An example of a density functional Hamiltonian that takes this density gradient into account is ***pBP***. Some computational results for SVWN (linear) and pBP (non-linear) computations are given in Tables 9-10 and 9-11.

9.7 COMPUTATIONAL STRATEGIES

The purpose of this section is to help a chemist choose an appropriate molecular mechanics or electronic structure computational strategy for solving a chemical problem of interest. The elements that go into making such a decision have to do with the reliability of the desired property needed and the most computationally efficient approach. The relative reliability of results from various methods for organic compounds is shown in Table 9-12. The relative reliabilities of various methods for inorganic compounds, organometallic compounds, and transition state structures are difficult to assess due to the lack of experimental data. The types of information that can be obtained from computations on molecules include equilibrium geometry, geometry of transition state structures, vibrational frequencies, and thermochemistry.

In terms of finding equilibrium geometries of compounds, even very simple computations such as semi-empirical and small basis set *ab initio* computational methods provide good geometries as compared to experiment. As a consequence, it is almost always advantageous to use these simple computations as a starting point in higher-level computations such as large basis set *ab initio*, CI, or MP2 computational methods. If equilibrium geometries are desired for large molecules or biopolymers, the molecular

Table 9-12. The comparative performances of molecular mechanics and electronic structure methods for organic compounds are shown for comparison. Table adopted from Wavefunction, Inc. with permission.

task	molecular mechanics	HF	MP2	semi-empirical	local DF	non-local DF
geometry	A/G	G	G	G	G	G
transition state geometry	-	G	G	G	G	G
conformation	G	A/G	G	P	A/G	G
thermochemistry	-	A/G	G	P	A/G	G

G = good A = acceptable P = poor A/G = acceptable to good mechanics techniques are the best choice. It is important to realize that all of the methods discussed in this chapter are for gas-phase molecules. There are no terms in the Hamiltonians for solvent effects. If the equilibrium geometry of a compound is desired in the presence of a solvent, there are practical semi-empirical models available such as ***AM1-SM2***. However, even at the semi-empirical level, solution phase computations are formidable.

An important question that needs to be asked is how accurate of an equilibrium geometry is needed. A very accurate result is needed for certain desired properties that are sensitive to the equilibrium geometry. Examples of properties that have a high degree of sensitivity to equilibrium geometry include dipole moments and vibrational frequency calculations. The model for calculating vibrational frequencies assumes that the first derivatives with respect to nuclear positions are rigorously equal to zero. Equilbrium geometries from high-level computations should be used in order to obtain these types of properties.

Transition state geometries are inherently more difficult to locate than equilibrium geometries of molecules. The potential energy surface along a transition state structure is somewhat flat rather than a steep minimum as found in an equilibrium geometry. As a result, small changes in energy for a transition state structure can result in large changes in geometry. Since transition states involve bond formation and breaking, low-level computations may not lead to acceptable results; however, it is best to start with a low-level computation (i.e. molecular mechanics or semi-empirical) as a starting point for a higher-level computation. The vibrational frequency for the transition state structure should be computed. The structure should yield only one imaginary frequency in the range of 500-2000 cm^{-1} that is typical of normal frequencies. Very small imaginary frequencies of <100 cm^{-1} probably do not correspond to the reaction coordinate of interest. An additional check that can be done is animation if the software being used will produce it. The animation can be used to see if the imaginary frequency smoothly connects the reactants to products.

In terms of thermochemistry, it is best to write reactions with the least number of bonds forming and breaking. If possible, the reaction of interest should be written in terms of ***isodesmic reactions*** (reactions where the reactants and products have the same number of each kind of formal chemical bond). An important question to ask is if an absolute energy is important or if a comparison between different chemical species will allow for a particular trend to be deduced. Though, as shown in Table 9-12 that semi-empirical computations in general yield poor thermochemical information, the thermochemical data obtained from these semi-empirical computations can be used successfully to deduce trends such as proton affinities and acidities. Since thermochemical properties do depend on equilibrium geometry, high-level computations in general are needed for absolute thermochemical information.

The computational strategy in general has a common theme. Start with low-level computations for a somewhat optimized equilibrium geometry and then re-submit the optimized geometry into a higher-level computation. The nomenclature used to describe a computational route is given as follows.

level/basis set//level/basis set

energy or property — geometry

The level corresponds to the type of computation used such as HF, MP2, AM1, and so on. The basis set corresponds to STO-3G, 3-21G, 6-31G*, and so on.

PROBLEMS AND EXERCISES

9.1) Make a plot of the σ and σ^* wavefunctions of a H_2^+ molecule as given in Equations 9-23 and 9-24. Relate the distances from the respective nuclei, r_A and r_B, in terms of a nuclear distance R. Make plots of the wavefunctions for different values of R from the value of 0.5 to 3.00 Å.

9.2) Write out the molecular orbitals that are formed from the atomic orbitals in the HF-SCF-LCAO approximation of nitrogen monoxide in a diagram like that shown for carbon monoxide in Figure 9-5.

9.3) Using simple MO theory, predict the bond order for the following molecules: a) S_2, b) S_2^+, c) N_2, d) N_2^+, e) F_2^-, and f) C_2.

9.4) In a HF-SCF-LCAO computation on methyl chloride, determine the number of functions in the following basis sets: a) minimum, b) 6-31G, c) 6-31G*, and d) 6-31G**.

9.5) Determine which of the following electronic structure computational methods can possibly yield a ground-state energy below the true ground-state energy: a) HF-SCF-LCAO, b) full CI, c) MP2, and d) pBP. Be sure to justify your answer.

9.6) Explicitly show that a product of two s-type GTO's one centered at r_A with an exponent of α_A and the other centered at r_B with an exponent α_B can be expressed as a single function centered between points A and B.

9.7) One method for obtaining heats of formation of compounds is to combine computational bond separation data with experimental data. Determine how you would determine the heat of formation of methyl hydrazine from the calculated bond separation energy for methyl hydrazine,

$$CH_3NHNH_2 + NH_3 \rightarrow CH_3NH_2 + NH_2NH_2,$$

and the experimental heats of formation data for ammonia, methylamine, and hydrazine. Is the above equation for the bond separation of methyl hydrazine isodesmic?

Appendix I

Table of Physical Constants

Speed of light	c	2.997925×10^{8} m s^{-1}
Elementary charge	e	1.60219×10^{-19} C
Planck's constant	h	6.62618×10^{-34} J s
	$\hbar = h/2\pi$	1.05459×10^{-34} J s
Boltzmann's constant	k	1.38066×10^{-23} J K^{-1}
Avogadro's constant	N_A	6.02205×10^{23} mol^{-1}
Electron rest mass	m_e	9.10953×10^{-31} kg
Proton rest mass	m_p	1.67265×10^{-27} kg
Neutron rest mass	m_n	1.67495×10^{-27} kg
Vacuum permittivity	ε_0	8.854188×10^{-12} J^{-1} C^{2} m^{-1}
	$4\pi\varepsilon_0$	1.112650×10^{-10} J^{-1} C^{2} m^{-1}
Bohr magneton	μ_B	9.27408×10^{-24} J T^{-1}
Bohr radius	a_0	5.29177×10^{-11} m
Rydberg constant	hcR_∞	2.179908×10^{-18} J
	R_∞	1.097373×10^{5} cm^{-1}

Appendix II

Table of Energy Conversion Factors

1 erg = 10^{-7} J

1 cal = 4.184 J

1 eV = 1.602177×10^{-19} J = 1.602177×10^{-12} erg

1 hartree = 4.35975×10^{-18} J = 27.2114 eV

1 rydberg = 1/2 hartree

Appendix III

Table of Common Operators

position	$\hat{q}$	q
time	$\hat{t}$	t
momentum	$\hat{p}$	$\dfrac{\hbar}{i}\dfrac{\partial}{\partial q}$
kinetic energy	$\hat{T}$	$\dfrac{-\hbar^2}{2\mu}\nabla^2$
del squared	∇^2	
	Cartesian coordinates	$\dfrac{\partial^2}{\partial x^2}+\dfrac{\partial^2}{\partial y^2}+\dfrac{\partial^2}{\partial z^2}$
	spherical coordinates	$\dfrac{1}{r}\left(\dfrac{\partial^2}{\partial r^2}\right)r+\left(\dfrac{1}{r^2}\right)\Lambda^2$
legendrian	Λ^2	$\left(\dfrac{1}{\sin^2\theta}\right)\left(\dfrac{\partial^2}{\partial\phi^2}\right)+\left(\dfrac{1}{\sin\theta}\right)\left(\dfrac{\partial}{\partial\theta}\right)\sin\theta\left(\dfrac{\partial}{\partial\theta}\right)$

Index